T0068800

Agriculture: A Very Short Introduction

VERY SHORT INTRODUCTIONS are for anyone wanting a stimulating and accessible way into a new subject. They are written by experts, and have been translated into more than 45 different languages.

The series began in 1995, and now covers a wide variety of topics in every discipline. The VSI library now contains over 500 volumes—a Very Short Introduction to everything from Psychology and Philosophy of Science to American History and Relativity—and continues to grow in every subject area.

Titles in the series include the following:

Paul Brassley and Richard Soffe

# AGRICULTURE

## A Very Short Introduction

OXFORD
UNIVERSITY PRESS

# OXFORD
UNIVERSITY PRESS

Great Clarendon Street, Oxford, OX2 6DP,
United Kingdom

Oxford University Press is a department of the University of Oxford.
It furthers the University's objective of excellence in research, scholarship,
and education by publishing worldwide. Oxford is a registered trade mark of
Oxford University Press in the UK and in certain other countries

First edition published in 2016
Impression: 7

Published in the United States of America by Oxford University Press
198 Madison Avenue, New York, NY 10016, United States of America

British Library Cataloguing in Publication Data
Data available

Library of Congress Control Number: 2015959268

ISBN 978-0-19-872596-1

Printed in Great Britain by
Ashford Colour Press Ltd., Gosport, Hampshire.

# Contents

# Acknowledgements

We have spent our working lives thinking and talking about agriculture with numerous acquaintances, students, colleagues, and friends, all of whom have affected our views in one way or another. For their helpful comments on the manuscript of this book we are especially grateful to Angela Brassley, Caroline Brassley, Charles Brassley, Charlie Clutterbuck, Alan Cooper, Steve Jarvis, Anita Jellings, Rob Parkinson, Alison Samuel, our editors at Oxford University Press, Andrea Keegan and Jenny Nugee, and our anonymous readers. For all her work on the figures and photographs, we are most grateful to Carrie Hickman, and we would like to thank Dan Harding for his careful, sensitive, and expert copy editing.

# List of illustrations

# List of tables

# Introduction

Consider an egg. Agriculture is about the getting a hen to lay the egg. It's also about producing the raw materials for other things that you might eat with it, such as bread and butter.

Where did the hen come from? Of course it came from another egg, but in this case one that was fertilized by a cockerel so that a chick developed inside it. Most commercial-scale egg producers in industrialized countries now buy their laying birds from specialist breeders who employ geneticists to develop more productive hens, each of which can now lay 300 eggs or more per year. This is roughly twice their output in 1950.

Having obtained his hen, the farmer then has to find somewhere for her to live. Hens are the domesticated descendants of Indian jungle fowl, and left to themselves they retain some of the behaviour of their wild ancestors, which would find somewhere secluded to lay their eggs, and roost in trees at night. People who keep hens in small numbers therefore often have hen houses with laying boxes containing nesting material and perches on which the hens can roost. At the opposite extreme, conventional battery cages used to have two or three hens standing in a small cage on a wire floor. This type of cage has been banned in the European Union (EU) since 2012. It has been replaced by 'furnished' cages which are much bigger, contain more hens, often between forty and eighty, and have nest boxes and dust-bathing areas. About half of the eggs produced in the UK now come from birds in cages, with the rest mostly produced by free-range flocks. In the US over 90 per cent of eggs are produced in battery cages, although one or two states—California, for example—have banned their use.

The farmer also has to ensure that the hen has food and water. Hens are natural scavengers. They will eat seeds, fruits, grass and other green leaves, worms, and insects, and they pick up small pieces of grit which go into their gizzards and help physically to break down their food. When the hens are kept in cages, and also in many free-range systems, the farmer has to provide all the nutrients that the hen will transform into eggs: carbohydrates, proteins, fats, minerals (especially calcium for the shells), and vitamins. All of these are combined into pellets or crumbs in a mill, which on large poultry operations may be on the farm, but is more often on a factory site, sometimes close to a seaport for ease of access to imported ingredients. The carbohydrates mostly come from cereals such as wheat, barley, or maize (known as 'corn' in the US), which have been grown in fields, some of which may be thousands of miles away. Similarly, soya beans, many of which are produced in Latin America, are one of the main sources of proteins and fats for European and North American hens. Thus the egg, like many animal products, is linked to arable farming, or the production of crops.

Arable farmers produce the cereals, protein and oil crops, root crops, such as potatoes or (in tropical countries) cassava and yams, and all the other plants and plant-derived raw materials that make up animal food and the great bulk of human food. To grow cereals, for example, they take the seed, which in industrialized and many developing countries is now a very sophisticated science-derived source of genetic material, and sow it into soil which has been prepared with the aid of powerful machinery and enriched by manures and fertilizers. The growing plant can be protected from attacks by fungi and insects, and from

2

competition by weeds, by the enormous variety of herbicides, fungicides, and insecticides that have emerged from the chemical industry over the last seventy years. Finally the crop is harvested by large machines which separate out the grain that will be used to feed people and chickens from the rest of the plant.

But while all that is happening in developed countries and in some countries of the South, at the same time there are still many countries in which the seedbed is prepared by farmers bending from the waist and using the same kind of hoe that their ancestors have been using for hundreds of years. These farmers might not be able to keep hens because a disease called fowl pest is endemic to their area and they cannot afford the few pence that it would cost to vaccinate them against it.

The agribusiness corporation producing corn and soya beans using enormous machines fifty miles south of Chicago, the woman with her hoe and her plot of cassava in Mozambique, the Chinese collective farm worker in the rice fields, and the German family with their part-time dairy farm and their day jobs in Munich are all engaged in agriculture. This *Very Short Introduction* sets out to identify the common features of their activities and the universally applicable principles that determine what they do, to explain why the differences between them exist, and to explore some of the controversies that arise from their activities. It is neither a criticism, nor a defence, of either science based farming methods or traditional farming practices. It recognizes the diversity of a world in which developed and developing agriculture can be found anywhere and everywhere.

# Chapter 1
## Soils and crops

### Soil

In May 1934 a wind storm lifted a black cloud of soil from the Great Plains and dumped twelve million tonnes of it on to the city of Chicago (see Figure 1). Two days later the storm reached the eastern seaboard and the dust drifted in through the windows of the White House in Washington. It was a salutary reminder that human life is dependent upon just a few inches of fragile topsoil, as important to farmers as sunlight. As a resource to anchor and feed the crops, it is a living medium containing thousands of living organisms. Across the world the type and depth of soil is fundamental to the crops that can be grown. The first part of this chapter looks at what soils are and how they differ, how they are classified, and how farmers manage them.

Soil is a complex mixture of air, water, minerals, and organic matter that has evolved, in many cases, over thousands of years. Typically, air may account for 25 per cent of the volume of a well-managed topsoil, and water another 25 per cent. The organic matter could take up about 5 per cent of the volume, so that the actual mineral fraction, varying from large stones to tiny clay particles, only comprises 45 per cent of the volume, although much more of the weight of the soil.

1. **The US Dust Bowl in the 1930s: a vivid image of wind-blown soil on the move.**

The type of soil found in any particular location depends upon the parent material, the climate, the topography of the site, the various organisms living in and on the soil, and time. Some parent materials are solid rocks such as granite, sandstone, chalk and limestone, or slate, whereas others are superficial deposits such as riverine alluvium, which is material that has been carried by a river and then deposited on its flood plain. Alluvial soils can be some of the most fertile in the world, whereas soils formed on sand dunes are often at the opposite end of the fertility spectrum. Loess, like sand dunes, is formed of material transported by the wind, but in this case involves much finer material. It covers extensive areas of North America and northern China, in the latter case being derived from the Gobi Desert. There are also peat soils, formed in wet conditions, glacial drift soils, formed on materials deposited by melting glaciers, marine clay soils, such as those in parts of the Netherlands, derived from material originally laid down under the sea, and volcanic ash soils.

What develops from these and other parent materials then depends upon the local climate, in particular the prevailing temperature and moisture, the organisms that come to live there, and the topography of the site, especially the slope. The greater the slope, the easier it is for repeated freezing and thawing, or torrential rain, to move material downhill, so gently sloping sites tend to have deeper soils than steep slopes. Increasing temperatures accelerate the weathering of minerals and the breakdown of organic matter, and the greater the soil moisture the more soluble elements such as calcium move down from the surface layers, so that wet tropical or semi-tropical soils are often quite acid whereas drier soils in areas of similar temperature are less so. The plants growing on the soil extract water and nutrients from it; when they die, under natural conditions, they become part of the decomposing surface litter, and following their breakdown the nutrients are again available for reabsorption.

In tropical forests the amount of litter deposited on the surface may be up to ten times that found in a northern pine forest, but such is the rate of decomposition in the warmer conditions that tropical soils may still have only low levels of organic matter. Some of the decomposition and mixing into the soil of plant debris is carried out by small animals such as earthworms, millipedes, springtails, mites, nematodes (eelworms), ants, and termites. The other decomposers are the fungi, bacteria, actinomycetes (which are unicellular like bacteria but produce mycelial threads like fungi), algae, and viruses. At the opposite end of the scale are the vertebrates such as rabbits, moles, prairie dogs, and the blind mole rat, all of which contribute to the mixing of soil layers by their burrowing activities. And of course farmers mix the soil when they cultivate it.

There is a cyclic process at work here, generally known as the carbon cycle, in which atmospheric carbon dioxide is fixed into plants (see the following section 'How plants grow'), which either die and are decomposed, or are consumed by humans and other animals and so partly fixed into their bodies, with the residue emerging as dung and urine to be decomposed in their turn and become part of the soil organic matter, some of which will be oxidized and so return to the reservoir of atmospheric carbon dioxide. The rate at which all this happens has clearly been affected by the spread of agriculture, because grazing animals remove growing plants and convert them both to dung and urine, but also to gaseous carbon dioxide and methane. It has also been affected, probably to a greater degree, by the exploitation of fossilized carbon in the form of coal and oil, which adds to atmospheric carbon dioxide levels.

All these factors contribute to the formation of different types of soil, as we shall see later, but for the individual farmer what probably matters most is the texture and structure of the soil. Texture is determined by the proportion of mineral particles of different sizes in the soil, as shown in Table 1, which shows the classification system employed in Britain.

Different soils will contain more or less of these various particles. One made up of 40 per cent sand, 40 per cent silt, and 20 per cent clay, for example, would be classified as a loam according to the US Department of Agriculture system, whereas one with equal quantities of all three would be a clay loam. Accurate determination of particle size distribution requires a complex process of chemical treatment, sieving, and sedimentation, but experienced soil scientists can make an effective enough assessment by a process known as 'spit and rub'. The soil is moistened between the fingers and thumb, and the feel of the sample gives an idea of the texture. Coarse sand feels gritty, fine sand slightly less so, silt makes the sample silky or slippery, and clay makes the sample sticky.

The texture of a soil affects how easy it is to work and how fertile it is. A very sandy soil will drain easily, but will also be subject to drought and not especially fertile. Those farming such soils often

**Table 1  Particle sizes and properties of soil minerals**

|  | Diameter (mm) | Characteristics |
|---|---|---|
| Gravel or stones | Greater than 2 | |
| Coarse sand | 0.2–2 | Beach sand, visible grains |
| Fine sand | 0.06–0.2 | Egg timer sand |
| Silt | 0.002–0.06 | Flour, only visible with a hand lens |
| Clay | Less than 0.002 | Putty, only visible through an electron microscope |

Source: R. J. Parkinson, 'Soil Management and Crop Nutrition', in R. J. Soffe (ed.), *The Agricultural Notebook*, 20th edn, p. 4. © 2003 by Blackwell Science Ltd, a Blackwell Publishing company.

say that they need a shower of rain every day and a shower of manure on Sundays, although they may not use those precise words. Clay soils, on the other hand, are much more fertile but much more difficult to work.

What is the reason for the difference? Sand and silt-sized soil particles have a relatively small surface area per gramme, and are largely composed of quartz (silicon dioxide) particles, which are chemically unreactive. Consequently they are less likely to retain water by capillary attraction and nutrients in chemical combination. Clay particles, on the other hand, have an enormous surface area for a given weight, often a thousand times greater than that of sand, and are chemically reactive, so they retain water and nutrients. Thus they are potentially more fertile, but on the other hand it requires more power to pull a plough or any other cultivator through a clay soil than a sandy soil, and clays are slow to drain and difficult to rewet in times of drought. Clay soils are often therefore described as 'cold' and 'heavy', whereas sandy soils are said to be 'light'. Most soils, of course, and certainly the best ones from a farming viewpoint, such as loams or clay loams, have neither too much sand nor too much clay and so are both fertile and easy to work.

Agriculture

Farmers can affect the water content of the soil, and so improve its workability, lengthen the growing season, improve the efficiency of fertilizer use, and encourage crops to root more deeply, by artificial drainage. The impact on crop yields in Britain has been found to be between 10 and 25 per cent. Various techniques are employed, from open ditches leading to streams and rivers, to tile or pipe drains and mole drains. The mole, in this case, is a pointed cylinder of about 70 mm diameter attached to the base of a vertical blade which is pulled through the soil by a powerful tractor. In clay soils the channels so produced may last for up to five years. Tile or pipe drains, laid by specialist machines, last far longer but are correspondingly more expensive to install.

The opposite process, of course, is irrigation, which has been carried out in some countries for hundreds if not thousands of years: the seasonal inundation of the Nile Delta is an obvious example. In temperate countries it is often only higher value crops such as potatoes, sugar beet, and vegetables that will repay the cost of irrigation, but in tropical areas irrigation is an integral part of the cultivation of paddy rice crops. One estimate suggests that 20 per cent of the global area of arable land, accounting for about 40 per cent of total crop production, is now irrigated. Much of this land is accounted for by the Asian rice crop, but there is a lot in the US too.

The other important consideration for the farmer is the structure of the soil, because that, too, affects how easy it is to work. The development of a soil, as opposed to a simple mixture of sand, silt, and clay, requires the mixture of the mineral particles discussed earlier with organic matter, by the activity of soil organisms over time. The result is the formation of stable aggregates that contain the water and air needed by plant roots and other soil organisms (see Figure 2).

The pore spaces can be filled with either water or air or a mixture of the two, and a well-structured soil has an even

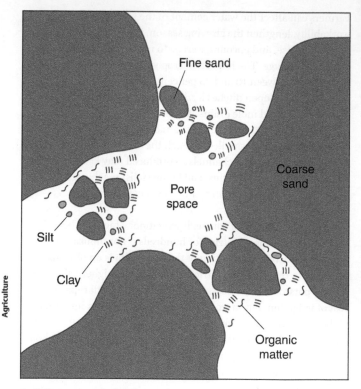

2. Soil structure in detail.

granular topsoil with gradually larger aggregates at greater
depth. It is relatively easy, in such soils, for cultivation
implements to produce the finely structured surface soil that
farmers call 'a good tilth', which provides the ideal conditions in
which plants can germinate and grow. In contrast, in a poorly
structured soil, such as a clay loam that has been worked when
too wet, the clay will have become 'puddled', or brought together
in large lumps, often with an impervious pan at 25–30 cm depth
which impedes drainage.

Farmers cannot change the texture (percentages of sand, silt, and clay) of their soil, but their farm management decisions can affect its structure. Continuous cropping—or, in the wrong circumstances, even regular arable cropping—can reduce the level of organic matter in the soil, making such structural problems more likely, whereas the use of permanent pasture, or at least long periods under grass, and regular additions of manure and lime, all help to improve soil organic matter and soil structure. In extreme circumstances the structure deteriorates to such an extent that it becomes susceptible to wind erosion, as was seen in East Anglia in the 1960s and most famously in the Dust Bowl of the Midwestern United States (see Figure 1).

All these variations in parent material, texture, structure, organic matter, and topography mean that there is an immense variety of soils in the world, for which soil scientists have developed various systems of classification, many of them involving extremely complex descriptive terms. The most widely used systems are related to climatic variables, mainly rainfall and temperature, and consequent vegetational zones.

Thus brown earths (sometimes called Cambisols) develop under temperate deciduous forests and make good mixed farming soils, and chernozems are the black soils that develop on steppes or prairies and form the arable farming soils of the US Midwest and the Ukraine. Acid, leached podzols develop in the cold and wet conditions found in coniferous forests, heaths, and moorlands, whereas in the warm wet conditions of the tropics, highly weathered lateritic soils (also called latosols or ferralsols), sometimes red or yellow in colour, are found. They are acid and nutrient deficient, and although they can support luxuriant forest sustained by rapid nutrient recycling at the surface, they form only poor agricultural soils. Gleys are waterlogged soils which in temperate regions need drainage and careful management, but in the tropics constitute some of the major rice-growing soils.

These are just a few of the wide range of soils recognized in climate-based classification systems, but there are other approaches which place greater emphasis on the significance of parent material and topography, and thus develop global classifications relating to geography and land form.

The soil, therefore, physically supports growing crops and supplies water to their roots. Dissolved in the water are the various nutrients that plants need to help them extract their main constituent, carbon, from the carbon dioxide in the air, and we shall look further at those in the following section.

## Crop nutrients

Plants get their major constituent elements—carbon, hydrogen, and oxygen—from the air they take in through their leaves and the water taken up by their roots. Apart from those, they require three elements in large quantities (macronutrients), another three in rather smaller amounts, and very small amounts of another eight, usually called the micronutrients.

The three major nutrients are nitrogen (N), phosphorus (P), and potassium (K). Most of the nitrogen is acquired from the soil water in the form of nitrate ions, and is a constituent of the plant's proteins and chlorophyll. Without it plants become yellow and fail to thrive. Ultimately, the nitrogen in plants comes from the air, but first it has to get into the soil. A small amount arrives dissolved in rainwater, but more comes from the activities of the soil-living algae or bacteria, such as *Azotobacter*, or from the nitrogen-fixing nodules on the roots of leguminous plants such as peas, beans, clovers, and even gorse. From these organic sources it is then converted by further bacterial activity to nitrate ions, which are very soluble in water, and in this form are taken into the plant roots. They may also, however, be leached out as water drains away, or converted into atmospheric nitrogen again, or taken up by soil organisms to once again become part of the organic reserve in the soil. What most manures and mineral fertilizers do is add to the amount available at the nitrate stage of the cycle.

Phosphorus is required for the energy and protein metabolism of plants, and without it roots do not grow well. There is usually plenty of it in the soil, but in a relatively insoluble form, so in some circumstances only the small amounts in solution may be available to plants. Potassium is involved in the water control and transport mechanisms of plants, so growth rates are reduced without it. It is released during the weathering of clays in the soil, and can be lost in drainage water as a result of its solubility. In normal circumstances and most farming systems, it is only these three major nutrients that are lost in sufficient quantities to require replacement by fertilizers.

There are, however, other elements that may sometimes need to be replaced in more intensive farm management systems. Sulphur is a constituent of some amino acids and enzymes, and in normal circumstances there are sufficient soil reserves, but since the decline of sulphur emissions from coal-fired power stations in Europe there have been signs of sulphur deficiencies in some areas. The other two elements in this category are calcium and magnesium, which play a part in controlling the acidity of the soil (see next paragraph). Finally, there are the micronutrients: boron, copper, zinc, manganese, molybdenum, iron, chlorine, and cobalt, many of which play important roles in plant enzymes but are usually required in such small quantities that most soils contain adequate amounts. For example, whereas a cereal crop producing 6 tonnes of grain and 3.5 tonnes of straw would remove 120 kilogrammes of nitrogen from a hectare of land, it would result in the loss of less than a kilogramme of zinc.

Soil acidity also affects the availability of nutrients. It is measured by the pH of the soil, a pH of 7 being neither acid nor alkaline, with the acidity increasing as the pH decreases. A cultivated soil normally has a pH between about 5 and 7, although on calcareous soils such as those on limestone and chalk the value could be between 7 and 8. The most acid upland peat soils might fall as low as 3.5. At this level of acidity the availability of all of the macronutrients and some of the micronutrients is reduced. In a slightly acid soil, with a pH between 6 and 7, nutrient availability tends to be at a maximum, as is micro-organism activity, so farmers recognize the need to maintain it at that level, by adding appropriate dressings of liming materials such as ground limestone. Lime, which also aids good soil structure, should therefore be seen as a soil conditioner rather than a fertilizer.

## Manures and fertilizers

From the preceding discussion, it is clear that the purpose of manures and fertilizers is to replace the plant nutrients that cropping and grazing remove, so they are mostly valued for the N, P, and K that they contain. There is no hard line between what constitutes a manure and what is a fertilizer. Essentially, manures come from organic sources—the classic example is farmyard manure, which is a mixture of straw, dung, and urine—and the feedstocks for fertilizers are produced in chemical factories (in the case of N fertilizers) or mined (in the case of P and K). But there are plenty of examples of products which are midway between these two extremes, such as blood and bone meal, or dried chicken droppings, which originate from an organic source but have then gone through a manufacturing process.

The nutrient content of manure is relatively low: even the most concentrated, which is poultry manure, only contains about 2 per cent of N and P and 1 per cent of K. The additional benefit of farmyard manure is that it contains partially decomposed organic matter which helps to improve the structure of the soil. In recent years, on highly mechanized farms, animal dung and urine have been collected as a liquid slurry, which is then applied directly to the land, without the addition of straw, so this will not produce the same soil structure benefits.

Organic fertilizers (as opposed to manures) are more concentrated, with hoof and horn meal and dried blood containing 12–14 per cent N. The range of such materials used in the past was enormous: guano, gas lime, fish, blubber, 'greaves' (candle-makers' waste), furriers' clippings, feathers, wool, linen rags, shoddy, fellmongers' poake (carcass remains), and rape dust all appeared on a list compiled in 1860. Similarly, many materials, from burnt lime to marl and shell sand, have been used in the past for their lime content.

Inorganic fertilizers are the most concentrated, with available nutrients accounting for up to half the weight of the material applied. There are several kinds of nitrogenous fertilizers, such as ammonium nitrate, urea, and anhydrous ammonia, but in each case the nitrogen content has been fixed from the atmosphere (a process which requires a lot of hydrocarbon fuel, now often in the form of gas) before being combined with other elements. Similarly, most phosphate fertilizers are derived from rock phosphate, which can be used directly as a fertilizer, but is very slow to release its phosphorus. Consequently it is usual to treat the rock phosphate with sulphuric acid (to make superphosphate) or phosphoric acid (to make triple superphosphate, with a higher P content).

Potassium is mined in the form of potassium chloride, which is either used directly or treated with sulphuric acid to make potassium sulphate. All these individual nutrients can be used on their own—in which case they are often called 'straights'—but they may also be combined to make 'compound' fertilizers which contain a mixture of nutrients, sometimes formulated to suit the requirements of particular crops. Thus a good general purpose compound fertilizer would be a 20;10;10, containing 20 per cent N and 10 per cent each of P and K, whereas a fertilizer designed to replace the nutrients removed by a cut of silage would require no P, so might be a 25;0;15 (i.e. 25 per cent N and 15 per cent K). How much of any of these to apply to a specific field or crop depends upon a variety of considerations, from the existing N, P, and K status of the soil to the response of the crop, its sale value, and the cost of the fertilizer. Most crops have a diminishing response curve, meaning that the additional crop yield response decreases as the level of fertilizer application increases.

The sustained production of crops and grass over long periods of time therefore requires healthy, well-managed soils. Some farmers believe that these can best be produced by avoiding the use of mineral fertilizers and relying on organic methods to recycle nutrients. Others use precision farming methods to maintain soil fertility by placing mineral fertilizers only where they are needed. Farmers have to make the right management decisions according to the resources available to them and the crops they wish to grow.

## Crops

Humans live on plants, either directly, by eating them, or indirectly, by eating the flesh or products, such as milk, that animals produce as a result of eating them. A few of the plants, such as blackberries or mongongo nuts, might be gathered from the wild, but most are cultivated, in which case they are called crops. Some sort of starchy energy crop forms the basis of human diets almost everywhere, often a grain crop, such as wheat, millet, rice, or maize (see Figure 3). Where conditions are unsuitable for these, roots and tubers such as cassava, yams, or potatoes are grown. The following sections of the chapter look at how plants grow, what prevents them from growing, and how we can improve their growth, before moving on to some of the major crops and cultivation systems.

3. The most important food grains: (a) wheat, (b) millet, (c) rice, (d) maize.

## How plants grow

How do farmers capture the energy in sunlight and make it available so that you, the reader, have the energy to turn the pages of this book? They employ plants containing chlorophyll, which is the stuff that makes plants green. It absorbs radiation which provides the energy to drive a series of biochemical reactions within the leaves which take carbon dioxide ($CO_2$) from the air and combine it with water taken up by the roots to make simple carbohydrate molecules—sugars—and release oxygen ($O_2$). The process is called photosynthesis and is summarized in the equation

$$6CO_2 + 6\,H_2O \xrightarrow{\text{light}} C_6H_{12}O_6 + 6O_2$$

and it is difficult to exaggerate its importance. Without it there would be no life as we know it on this planet. It happens in all green plants, from the shortest grass to the tallest trees, and it is what produces 95 per cent of their dry matter (i.e. the part that isn't water). The simple sugars such as glucose can combine together in long chains (called polymers) to form starch, which is the form in which most plants store their energy, and cellulose, which is one of the main constituents of the plant cell walls. Fats and oils can also be produced by further transformations of carbohydrates, and, with the addition of nitrogen and sometimes other elements, proteins too.

Clearly, therefore, the speed at which the crop grows depends upon the rate of photosynthesis, and that in turn depends upon light, temperature, the carbon dioxide concentration and the availability of water and nutrients. As the temperature and light intensity increase so does the rate of photosynthesis, as long as water is available. For most temperate crops, which use what is called the $C_3$ metabolic pathway for carbon fixation, the limiting factor is the $CO_2$ concentration in the atmosphere. This is why increasing levels of this gas due to the burning of fossil fuels have a fertilizing effect. It is also why some growers enhance the $CO_2$ levels in glasshouses.

Many tropical crops such as maize, sugar cane, millet, and sorghum use the alternative $C_4$ metabolic pathway, in which photosynthesis is not limited by the $CO_2$ concentration until much higher light intensities. Plants with this metabolism can have high photosynthetic rates in hotter drier conditions, so research is currently going on to identify the genes responsible and to use them to produce a $C_4$ variety of rice. It has been claimed that this could yield up to 50 per cent more grain than current $C_3$ varieties.

As a plant grows, the dry matter or total biomass increases because the number of cells increases. The plant progresses from seed germination through vegetative development, reproductive development, and seed development by means of cell specialization (differentiation) and organization. Depending on its life cycle (annual, biennial, or perennial) and the part of the plant that is useful and harvested as food, a crop may reach senescence (death) after weeks, months, or years.

Before germination most seeds are resistant to cold and drought stress and can often survive for long periods. Those in the Svalbard Global Seed Bank, for example, are kept in a disused mine at below-zero temperatures in the permafrost. Most temperate crops need a temperature of at least 1–5°C before they will germinate at all, with optimum germination at between 20–25°C, but the optimum for rice is 30–35°C. In the vegetative stage the growing point(s) (meristems) of the plant produce leaf and stem material, and in the reproductive stage flowers, which, following pollination and fertilization, produce seeds. For grain crops, farmers only want enough stem and leaf development to maximize the photosynthetic (green) area of the plant, because it is the seed that they will harvest and sell, whereas they do not want potatoes to produce seed at all, because it is the tubers that are harvested as food.

Most crops grow faster as the temperature increases, and crops such as maize that originated in warmer climates normally require higher temperatures for optimum growth than those such as wheat or ryegrass that were originally found in cooler parts of the world. Temperature drives the rate of development, and often controls the timing of the change from one stage of growth to another. For example, some varieties of wheat are designed to be sown in the autumn (winter wheat), and require a period of cold weather (vernalization) in order to set seed in the following summer. Some plants also respond to day length. Some rice varieties, for example, will not flower until the days begin to get shorter, whereas most temperate plants are 'long-day' types, needing increasing day length to begin flowering, which ensures that they do so in the summer.

All these are features of the individual plants, but the crop is a collection of plants, and the interactions of the individuals are also important. The farmer seeks to manage the crop as a whole to maximize the amount of light intercepted at the time when the days are longest and the sun is at its most powerful. Thus if it is sown too late it may not have produced enough leaves to cover the ground, and sowing too thinly will have the same result. A good rich soil and enough water will also encourage leaf growth. On the other hand, take any of these factors too far and the result will be a decrease in yield, as higher leaves shade out the lower ones, individual plants have insufficient room in which to grow, or rapid growth or water shortages produce weak or floppy stems that do not remain erect.

A balance has to be struck between the yield of the individual plant and of the crop as a whole. The skill and experience of the farmer and the work of the agricultural scientist are needed to determine optimum seed and fertilizer rates and sowing times in order to maximize the output of the crop.

## What prevents plants from growing?

It follows from these characteristics of plants and crops that anything that prevents individual plants and the crop as a whole from operating at their maximum photosynthetic efficiency will reduce yields and outputs. In theory, a typical crop could capture about 5 per cent of the sun's energy available to it; in practice the figure is closer to 1 per cent, partly as a result of climatic factors, such as cold or cloudy conditions, or through a shortage of water or nutrients, but mainly due to the effects of weeds, pests, and diseases.

A weed is proverbially a plant in the wrong place. Poppies (Papaver spp.) look pretty in the garden but are a problem in a wheat crop. The grass *Imperata cylindrica*, known as *lalang* in Malay, *cogon* in the US, and *blady grass* in Australia, is a traditional thatching material in South East Asia, and also cultivated for beach stabilization, and sometimes grazed in Africa, but is subject to official eradication schemes in several states of the south-eastern US. The significance of a weed can also depend on the time at which it grows: whereas an early weed infestation can smother a crop, late weed establishment in a fast-growing crop may have virtually no effect on yield. Thus the most obvious impact of weeds arises from their ability to compete for sunlight and so reduce crop photosynthesis, but they also compete with the crop for water, their seeds may contaminate the harvested crop, and they may make harvesting difficult, as for example when bindweed climbs up crop stems.

Almost any plant may therefore be a weed if it grows in the wrong place at the wrong time, but some plants are recognized as especially likely to cause problems. In temperate crops particular difficulties are caused by grass weeds, such as blackgrass and wild oats, annual broadleaved weeds, and perennial weeds. Fat Hen (*Chenopodium album*), for example, is now normally classified as an annual broadleaved weed, although in the past it has been used as a salad vegetable. The perennials include couch grass, bracken, docks, and thistles in temperate areas; in tropical areas one of the most troublesome weeds has been water hyacinth (*Eichhornia crassipes*), which was introduced from Brazil as an ornamental plant, escaped, and flourished to the extent that it blocked irrigation channels and caused floods.

Like weeds, pests are animals in the wrong place. The elephant that looks so magnificent on the plains of the Serengeti is a pest when it tramples through a crop; the rabbit may be welcome to nibble the grass in a woodland glade but not the young wheat on the other side of the hedge. At the opposite end of the size scale are minute eelworms (properly called nematodes) which can damage sugar beet and potatoes, and numerous insects, including aphids and locusts. They can attack all parts of the crop from the seeds to the roots. Some affect the plant's ability to photosynthesize by eating its leaves, and others its source of water and nutrients by eating the roots. Aphids have piercing mouthparts and feed on the sap in the plant cells, so stunting its growth, and in doing so some of them, such as the grain aphid (*Sitobion avenae*), pass on virus diseases (barley yellow dwarf virus or BYDV in this case) which further attack the plant.

Some of the most devastating pest attacks have happened when the causal organism has been introduced into a new environment. In the US the Colorado beetle (*Leptinotarsa decemlineata*) lived a quiet life on weeds of the nightshade family until about the middle of the 19th century. Then potatoes, which belong to the same botanical family as the nightshades (the *Solanaceae*) were introduced, and the population increased and spread. From Colorado, it reached the Atlantic coast of the US in 1873, crossed the Atlantic in 1901, and from there spread to continental Europe as a major pest of potatoes. Similarly the phylloxera aphid, to which American vines were relatively immune, caused havoc when it spread to continental European vineyards in the 19th century because the vine varieties used there, never having been exposed to it, had not evolved the same protective mechanisms as American varieties.

The range of insect pests is enormous, with almost every crop having several species that will live on it if environmental conditions are appropriate. African coffee crops, for example, are subject to the attacks of the variegated coffee bug, a capsid bug, the berry borer, mealy bugs, scale insects, stem borers, and leaf miners. Over the world as a whole, one estimate puts the loss of crops due to pests at as much as 25 per cent.

Diseases of crops are mostly caused by various species of fungi, but viruses and bacteria can also be responsible. Some fungi cause a loss of chlorophyll, making the leaf look yellow, and so reduce photosynthesis. Others grow in the vessels through which the plants transport water, making them wilt. An example of such a fungus which is currently causing considerable problems is *Fusarium oxysporum*, which lives in the soil and penetrates the roots of banana plants, eventually spreading so that the leaves turn yellow and the whole plant begins to wilt.

*F.oxysporum* was first identified in Panama at the end of the 19th century, and so is also known as 'Panama disease'. The variety that dominated the commercial banana trade at that time, 'Gros Michel', was susceptible to the disease, which by 1960 had spread to most producing countries. Fortunately, another variety, 'Cavendish', proved resistant to the races of *F.oxysporum* then existing, so it in turn became the dominant commercial variety, but in recent years a new race, termed Tropical Race 4 (TR4), has emerged from Malaysia and spread to other countries, and 'Cavendish' is susceptible to it. Potato blight, caused by the fungus *Phytophthora infestans*, is perhaps the best-known example of a plant disease that had an enormous socio-economic impact when it attacked the potato crop in Ireland in the middle of the 19th century.

## How can we improve plant growth?

If plants are prevented from growing by shortages of light, heat, water, and nutrients, the obvious solution is to provide these things, and it is often possible to do that, but at a price. If the crop is valuable enough, the price may be worth paying. Some rich English landowners in the 18th century erected heated glasshouses in which to grow pineapples, and more recently police raids have found completely artificial growth rooms in which drug dealers have been growing cannabis plants.

In more common commercial circumstances tomato growers in northern Europe grow the crop in glasshouses, artificially heated, watered, fertilized, and in some cases with increased $CO_2$ levels, whereas in southern Spain the same crop needs no artificial heat but requires much more water to be provided from underground boreholes. Air travellers over dry parts of California can look down and see green circles in the desert that show where crops are being grown with the aid of large irrigation rigs circulating round a central water source (see Figure 4).

If transport costs can be kept low enough, it may be worth growing a crop where it is warm, for sale where it is too cold for it to grow: hence the Kenyan green bean in European supermarkets in winter months. Farmers everywhere have always tried to return organic nutrients to the land in the form of animal manure, and most farmers (except organic farmers) in developed countries, and increasing numbers in developing countries, now use artificial fertilizers (see earlier in this chapter).

Even in ideal growing conditions, crop growth may still be reduced, as we have seen, by the effects of weeds, pests, and diseases, and part of the skill set of the farmer is knowing how to reduce their impact. Weeds are most difficult to control, either by cultivation or

4. Centre pivot irrigation promotes crop growth in dry California.

herbicides, when they are very similar to the crop in which they are growing. If the weed looks like the crop one hesitates to attack it with a hoe, and if it resembles the crop physiologically and anatomically it will be difficult to find a herbicide that can kill it without also damaging the crop. Wild red rice (*Oryza rufipogon*), for example, is the most serious weed of rice, with losses in the Punjab at one time being estimated at over half of the crop.

Farmers have developed numerous strategies for weed control. It is important to sow uncontaminated seed at the right seed rate into a fertile seedbed to give the crop the best chance of competing with weeds; some weeds can be controlled by repeated cutting; and rotating crops enables broadleaved weeds that were difficult to control in a broadleaved crop to be tackled more effectively in a following cereal crop, and vice versa.

Before the introduction of chemical herbicides farmers and their workers would go through a crop with hoes, or pull out weeds by hand, and for high-value seed crops with fairly light weed infestations this is still sometimes done today. One of the reasons for inventing the seed drill, which planted seeds in rows, was that it was then easier to hoe between the rows with a horse-drawn hoe, as Jethro Tull argued in his book *Horse-Hoeing Husbandry* (1733), and crops with widely spaced rows, in which it is easier to employ inter-row cultivations, are often called 'cleaning crops'. But all these methods need a lot of labour.

At the beginning of the 20th century, before chemical herbicides were available, one farmer estimated that between a third and a half of the field labour on a farm was devoted to the destruction of weeds. It is therefore easy to see why farmers, especially in high-wage economies, have been such enthusiastic adopters of herbicides, and why organic farmers, who cannot use them, often argue that weeds are more difficult to manage than pests and diseases.

As with weeds, so with pests and diseases: farmers try to use a mixture of management and chemical methods to reduce their impact. Rotating crops, so that different ones grow on a field in succeeding years, helps to break the life cycle of fungi and soil-borne insects, and sowing early or late to miss the time at which the pest or disease is most active, are examples of management methods, and farmers increasingly resort to expensive agrochemicals only when these methods are ineffective.

Chemical control is itself not without problems. Not only may the target organism develop resistance to it, but it may also have unwanted side effects on non-target species. The impact of extensive use of chlorinated hydrocarbon insecticides on the breeding success of birds of prey is one well-known example. One solution to this problem is to use predators to attack pest species, as in the use of parasitic wasps to attack glasshouse whitefly; another is to attempt to breed resistance to attack into the plant.

Pest and disease resistance is just one of the possible objectives of the plant breeder. Essentially, most breeding programmes attempt to increase crop yield, or quality, or both, although sometimes increased yield is obtained at the expense of quality: for example, IR8, one of the early high-yielding rice varieties, released in 1966 by the International Rice Research Institute in the Philippines, was disliked by consumers because the grain was chalky and brittle.

There are various ways in which breeders can attempt to increase yield. They can produce varieties that are resistant to drought or to common pests and diseases, such as diseases of cereal foliage. They can also try to increase the saleable part of the plant at the expense of the rest. This has been extensively used by both rice and wheat breeders, who have produced varieties with shorter straw but bigger ears. The total biomass may not change much, but more of it can be sold.

Other objectives depend on the particular characteristics of individual crops. Older sugar beet varieties produced seed in clusters, which would produce three or four plants when they germinated. These had to be laboriously singled by hand hoe to produce one strong plant, so when breeders produced 'monogerm' seed, with only one seed per cluster, the need for this job was removed and costs were correspondingly reduced. Some varieties of the other source of sugar, sugar cane, have sharp leaf edges, which can be painful for workers in the cane fields, so breeders have attempted to produce varieties which minimize this characteristic. Modern plant breeding methods and their implications are discussed in Chapter 6.

## Crops and cropping

'Most of us have a potato-shaped space inside that must be filled at every meal, if not by potatoes then by something equally filling' wrote Katharine Whitehorn in her classic *Cooking in a Bedsitter* (1963). These staple foods are usually starch-based, and come from some of the most widely grown crops, such as rice, wheat, maize, potatoes, cassava, and yams. The science fiction writer John Christopher, in his novel *The Death of Grass* (1956), suggested that world civilization would break down if a disease began to kill off all the *Poaceae*, the family of plants that includes not only the grasses, which are the most widely grown crop in the world, but also the temperate cereals (wheat, barley, oats, rye), the tropical and semi-tropical cereals (maize, rice, the millets, sorghum), and sugar cane. We should also be short of proteins if a similar plague affected the legumes (*Fabaceae*), the family that includes the peas, beans, and lentils, and of vegetables and oilseeds if the *Brassicaceae* (cabbages, rape, swedes, mustards) were attacked.

Some crops produce more than one product. Wheat makes bread and barley is malted for beer, but both are also used to feed animals, especially pigs and chickens. Soybeans can be crushed to produce oil, and what remains is then a high-protein animal feed.

27

Linseed stems are the source of flax fibre to make linen, while the seeds are the source of an industrial oil, although over time different varieties of the species have been developed to specialize in one or the other product.

There are different varieties of most crops, hundreds of different varieties in the case of wheat, for example. Some have been bred to grow better in hotter and drier climates, some to produce better breadmaking flour, others to be resistant to various fungal diseases, and so on. The varieties of maize that are grown in tropical and semi-tropical countries and form one of the most important sources of dietary starch would not be those grown to produce maize silage for animal feed by farmers in north-west Europe, or to produce sweetcorn by American farmers. The grape varieties that produce the great wines are not designed to be eaten as fresh table grapes.

These are all human or animal food crops, but there are also extensive areas of crops such as cotton, jute, and hemp which are grown for their fibres, and smaller areas of borage, calendula, and evening primrose for industrial and pharmaceutical oils. Increasingly, crops may be grown to produce biomass, or fuel such as bioethanol.

All these species may also be classified according to their mode of growth. There are tree crops, such as coffee, cocoa, bananas, olives, and apples which, once established, grow for many years. Other perennials may be bush crops, such as grape vines, currants, or raspberries, which again can remain in place, for many years in the case of vines. But the majority of crops that farmers grow are annuals or (a few) biennials. In temperate climates that means the crop usually grows during the summer and is harvested in the autumn.

The farmer prepares a seedbed, sometimes by ploughing to bury the remains of the previous crop, then cultivating to break down the clods of soil, before depositing the seeds in rows using a seed drill. Some cereals such as wheat or barley may be sown in the autumn and become established before the really cold weather occurs. They are then ready to grow when the soil warms up in the spring, and so maximize the period when they can photosynthesize most effectively. The time available to prepare a seedbed in the autumn may be limited, or winters so harsh that many seed crop varieties are sown in the spring. In northern Canada, for instance, crops may have only ninety days in which to complete the cycle from sowing to harvest. Root crops such as potatoes and sugar beet are normally planted in spring for harvest in the following autumn. Once the crop is established it may be given one or more dressings of fertilizer, and herbicide and fungicide sprays. Harvesting takes place when the grain has filled and matured, in the case of cereals and oilseeds, and when little further growth is expected in the case of potatoes and sugar beet.

This is obviously a very basic outline of the process, and it should be clear that it involves many different decisions, about the best way to prepare the seedbed, which varieties to plant, what seed rate to use to produce the optimum plant population, how much fertilizer to use and when to apply it, which weeds, pests, and diseases might threaten the crop and how to combat them, when to harvest, and so on, all of which may vary from farm to farm, field to field, and even within the same field.

Farmers also have to decide which combination of crops to grow. Sometimes they grow one crop year after year, but this can lead to a build-up of pests and diseases, which have to be controlled, and a depletion of soil nutrients, which have to be replaced. There is a balance to be struck between growing the most profitable crop, and maintaining crop health and soil fertility.

A modern six-course arable rotation in Europe might see two years of wheat followed by one of beans, followed by wheat, then barley, then oilseed rape (canola). The beans break the wheat pest and disease life cycles and, being legumes, help to fix nitrogen in the soil; the oilseed rape also breaks the cereal disease life cycle. The classic Norfolk four-course rotation had wheat followed by turnips, which were heavily dressed with farmyard manure and also carefully weeded, so that there was a rich and weed-free soil for the following barley crop, which was then followed by clover, which fixed nitrogen and so enriched the soil for the following wheat crop, and so on. Modern fertilizers and pesticides have enabled farmers to increase the time between the restorative crops, but the principle remains applicable, and cheaper than complete reliance on chemicals. Rotations are simplified rather than eliminated.

In the Midwestern states of the US, a region often called the Corn Belt, farmers now rely on a rotation of corn (confusingly called maize in Britain, where 'corn' is a collective noun for wheat, barley, and oats) and soya beans. In the southern states, corn/maize may be grown in rotation with cotton and a legume such as cowpeas. A variation on this theme may be found in some tropical countries, where a crop such as cassava, which stays in the ground for several years, may either be grown in a pure stand or intercropped—grown in association with—annual crops. In some rice-growing areas the land may remain fallow until the next rice crop is sown, but in some parts of South East Asia, where the soil is suitable for cultivation in the dry season, the rice may be rotated with pulses, maize, or vegetables.

When farmers decide on the best crops to put in a rotation, they also have to think about the ways in which they can harvest them: all the crops in the wheat/beans/barley/rape rotation mentioned earlier can be combine harvested, whereas inserting a grass break would require the farm to have grazing animals, or additional machinery to cut the grass and make it into hay or silage.

The decisions that farmers make about the crops they grow therefore depend upon a wide range of factors: not only what the soil and climate will allow them to grow, but also what they have the knowledge and equipment to produce, and what they or their animals need to eat or can sell. These considerations are discussed in the following chapters.

# Chapter 2
# Farm animals

Humans have domesticated a relatively small proportion of the available animal species. As the American polymath Jared Diamond has argued, animals needed particular characteristics to be suitable for domestication: the right diet, usually based on a wide range of plants, a relatively rapid growth rate, an ability to breed in captivity, and a suitable disposition and social organization, so that they don't panic and have a well-developed dominance hierarchy in which humans can take over as leaders. Only a few species combined all of these characteristics, and they have become the most common farm animals. The top ten (i.e. the most commonly kept) farm animals are shown in Table 2.

These numbers need to be used with care, because obviously one chicken produces nothing like the same amount of food as a cow, a sheep, or any of the bigger animals. They also need to be set in a time context. The number of chickens and ducks in the world roughly doubled in the twenty years since 1992, and goat numbers increased by 65 per cent, whereas cattle and pig numbers only went up by about 13 per cent, and sheep numbers were static. Furthermore, farmers keep other animals that do not appear in this list, such as the llamas, alpacas, and guinea pigs of South America, yaks, and even insects such as silk moths and honeybees.

**Table 2  World farm animal population 2012 (million head)**

| Chickens | 21,867 | Pigs | 966 |
|---|---|---|---|
| Cattle | 1,485 | Turkeys | 476 |
| Ducks | 1,316 | Geese and guineafowl | 382 |
| Sheep | 1,169 | Buffalo | 199 |
| Goats | 996 | Camels | 27 |

FAO.FAOSTAT. Live Animals: Number of Heads (FAO 2015) Accessed 29 July 2014
<http://faostat3.fao.org/faostat-gateway/go/to/browse/Q/QA/E>.

Some animals, such as reindeer, muskoxen, or ostriches, are best described as semi-domesticated, as they are exploited by people but are not managed as intensively as most farm livestock. Others, such as horses and donkeys, are used in farming, but also in non-agricultural transport roles, as are cattle in some parts of the world, so that it is difficult, and not always useful, to measure the numbers of livestock kept for purely agricultural purposes.

Many farm animals have multiple functions. Cattle produce milk (and blood in some parts of Africa) and in some parts of the world pull ploughs and carts while they are alive, and produce meat and leather when they are dead; sheep are shorn for their wool and produce milk as well as being slaughtered for meat, and camels and buffaloes produce milk and pull ploughs.

It is perfectly possible to live on a vegan diet, with no animal products at all, and in developed countries we no longer rely on animal fibres for clothing or animal power for traction, so why do we bother to keep animals at all? Is it simply that animals produce a convenient collection of nutrients that are not always easily available to vegans? In part at least, it must be a question of tradition. Before the invention of steam and internal combustion engines, many animals—horses, donkeys, mules, cattle, and even dogs—were kept for their muscle power, to pull carts, ploughs,

Farm animals

33

and other agricultural implements. Their milk and meat were a by-product for which people developed a taste. Pigs and chickens, to which this argument clearly does not apply, were valued as scavengers that would provide meat and eggs by eating food waste or things like acorns or insects that were unpalatable to most humans. Thus, over the course of thousands of years of living with domesticated animals, humans have developed a taste (and are willing to pay) for the meat, eggs, and dairy products that they produce, and have found innumerable ways of combining them together and with plant products to make the food we eat. Farmers are therefore interested in making their animals grow quickly, and put on the right combination of muscle and fat that butchers prefer, or produce lots of milk or eggs or wool, and this chapter is about what they need to know in order to do that.

## Feeding farm animals

Some basic principles hold true for all kinds of animal husbandry (as the care of farm animals is often called). Boiled down to its essentials, and forgetting for a moment that farm animals are sentient creatures, with all that implies, animal husbandry is about converting the carbohydrates, proteins, fats, etc. in plants into the carbohydrates, proteins, fats, etc. in animals, so that they can give us meat and milk, and, in some cases, have the energy to pull our farm implements.

Animals possess the enzymes (biochemical catalysts) needed to break down the starch in plants into its component glucose molecules. Glucose is used directly to provide energy for the animal, or synthesized into glycogen for energy storage. In the mammary gland, it is combined with other molecules to form lactose, the main carbohydrate in milk. Mammals and birds do not, however, possess the enzymes necessary to break down the cellulose (dietary fibre) in plant tissues into glucose, although some bacteria and fungi do. This means that those animals that have evolved to live with a gut flora containing the right sort of

34

bacteria can extract nutrients from relatively fibrous leaves, containing a high proportion of cellulose. Those that have not must eat seeds and tubers containing the starch that they can metabolize.

This is a vital difference, because it means that the cellulose metabolizers (such as cows, sheep, goats, camels, and buffalo) do not have to compete with the starch metabolizers (such as chickens, pigs, and people) for food. In practice, of course, the carbohydrates in the diets of both groups include a mixture of sugars, starch, and cellulose, but what happens to them depends upon the gut anatomy of the various animals.

An animal's muscles, skin, hair, or wool, and even part of its bones, are made of proteins, which are large molecules formed by linking simpler compounds, called amino acids, of which there are about twenty different examples. Plants and bacteria can synthesize all twenty, but there are ten of them that cannot be synthesized by animals. These are known as essential amino acids, and they must be provided in the diet. Put simply, the digestive metabolism of the animal involves breaking down the proteins in its food into their constituent amino acids, which are then available to form the different proteins in the animal itself.

The other major constituents of the animal's body are the various fats, which are not only deposited under the skin, but are also found within and between muscles and muscle fibres, and also form parts of nerves, cell membranes, and hormones. Again, most of these can be synthesized by the animal from the fats and oils in its food. There are also minor nutrients, in terms of weight, which are nonetheless vital components of an animal's food. These are the minerals and vitamins, the lack of which prevents efficient metabolism and is usually demonstrated as disease, such as bone deformations (rickets) caused by vitamin D deficiency.

What happens to food after it has been eaten depends upon the anatomy of the animal's digestive tract. Simple-stomached (monogastric) animals begin the process of chemical breakdown of food in the stomach, after which the semi-digested material passes into the small intestine, where most of the remaining digestion of carbohydrates, proteins, and lipids occurs. What remains is then moved by muscular contractions into the large intestine, where water is absorbed before the undigested waste is voided in the process of defecation. This is in essence what happens in humans, who, as mentioned earlier, have little ability to digest cellulose.

The digestive tract of the pig is similar, except that the caecum (the part of the gut at the beginning of the large intestine) is larger than in humans and has a microbial population that can break down cellulose to a limited extent, so that up to 20 per cent of the pig's energy may be derived from this source. This means that pigs can derive some food value from grass and other vegetable fibres that would pass straight through a human.

The digestive tract of the chicken is completely different in detail from that of monogastric mammals up to the small intestine, but from a farmer's viewpoint the result is much the same: it needs a relatively low-fibre diet. In horses and donkeys, however, the caecum is much enlarged, and its microbial population has a considerable ability to break down cellulose. Roughages such as grass and hay can therefore form a large part of equine diets.

Ruminants, such as cattle, sheep, goats, and buffalo (and, with a different anatomy, camels), also carry bacteria that can break down cellulose, but in an enlarged and complex stomach called the rumen. After eating, ruminants can bring food back into their mouths and chew it again—the process of rumination, or chewing the cud—which helps to break down fibrous food for further metabolism by the rumen bacteria. In fact, without some dietary fibre, ruminant digestion does not work properly, which means that although highly productive cattle, for example, may be given concentrated foods containing starch, proteins, and fats, they also need fibrous material if they are not to suffer digestive problems. Furthermore, in milking cows high-concentrate/low-forage diets can depress the fat level in the milk.

These differences between ruminants and monogastrics are important from a farmer's viewpoint, because they mean that ruminants can live and thrive on a high-fibre diet of plant leaves on which a monogastric animal would struggle to survive. Consequently, they can utilize grass, which makes a good break crop on arable land (see Chapter 1), and can also exploit the vegetation on land that is too steep, rocky, or poor to be cultivated.

The way in which a farmer feeds animals therefore depends on their species and consequent anatomy and physiology, the feed available, and the purpose for which they are being kept. All animals have a 'maintenance' requirement for the nutrients necessary to maintain their body weight and enable their various physiological functions, without which they will eventually starve and die. Most of them also have a 'production' requirement, for growth, pregnancy, lactation, and activity.

In animals that are to be slaughtered for meat, feeding for growth is usually the major cost of production. In the young animal it is the bones that reach their peak growth rate first. As the animal ages, and the relative growth rate of bone decreases, it is the muscle growth rate that peaks, and finally, as muscle growth begins to slow down, fat deposition reaches a maximum (see Figure 5). Higher levels of fat are generally found in older or more rapidly growing animals within a breed.

Farmers try to sell their animals for slaughter when their fat levels are at an optimum: too little fat, and their meat is likely to be dry and flavourless; too much, and the animal loses value. But what constitutes too much can change over time. In the 19th century,

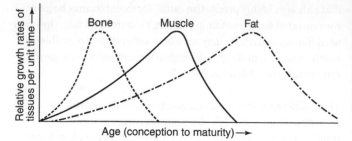

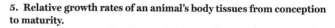

Age (conception to maturity) →

**5. Relative growth rates of an animal's body tissues from conception to maturity.**

for example, pigs were often fed to much greater weights than they would be today, because there was then a bigger market for their lard, or subcutaneous fat.

When an animal becomes pregnant, her production requirement obviously increases because in addition to maintaining her own body she also has to cope with the requirements of her growing offspring. Although this ceases after its birth, she then has to produce milk to feed it, which again needs additional nutrients over and above her maintenance requirements. In practice, both pregnancy and lactation may occur at the same time in dairy cows, because they become capable of conceiving again two or three months after giving birth, whereas their lactation can last for nine or ten months. In a high-yielding dairy herd the farmer will try to get the cows to have a calf every year, so that as the milk yield declines from its peak, the requirements of the growing embryo increase. High-yielding cows may need three times their maintenance requirement or more for these production requirements.

It is difficult to get enough nutrients into cows that are producing a lot of milk using forage alone, so they are usually given an additional ration of high-energy and protein concentrates. Even so, many high-producing cows will still not receive sufficient nutrients for their needs and will utilize backfat reserves. On the other hand, ewes on fresh spring grass, readily digestible and with a high protein content, may be able to feed their lambs without much if any additional concentrate feeding.

The corresponding situation for chickens is egg production. Although most hens could probably find their maintenance ration by foraging around a farmyard, a hen in a high-yield production system laying an egg almost every day needs a diet with a high nutrient concentration. Similarly, a horse needs a feed of oats or oats and beans if it is to pull a plough all day, whereas it needs little more than grass if it is not working.

## Breeding farm animals

In addition to feeding their animals for maximum performance, farmers can also affect their output by the way in which they breed them. It is important to distinguish between a breed and a species. Essentially, species cannot interbreed: the joke about producing woolly jumpers by crossing a sheep and a kangaroo is a biological impossibility. But within the same species animals of very different appearance can be successfully crossed. Fertilizing a female horse (mare) by a male donkey produces the mule which has been used as a pack and traction animal for centuries, and Indian humped zebu cattle (*Bos indicus*) can be successfully crossed with European-type cattle (*Bos taurus*). In the US, for example, at the beginning of the 20th century, breeders crossed Brahmans, a zebu breed, with Angus cattle originating in Scotland, to produce the Brangus beef breed. There have even been crosses of *Bos taurus* with the North American buffalo or bison (*Bos bison*) to form a hybrid known as the cattalo or beefalo.

Within a species, therefore, one breed is distinguished from another by differences in colour, size, and/or shape which are maintained from generation to generation. These differences probably originated in geographical variations, when farmers in one region produced animals which suited their particular local resources and needs. As communications improved, farmers elsewhere realized that they could improve their own stock by using those from another area. In the 18th century, for example, farmers in the north-east of England imported bulls from the Netherlands which 'did much service' in improving local cattle. An early 20th-century survey of British livestock listed thirteen major breeds of cattle, eight of pigs, and twenty-nine of sheep, in addition to various crosses and minor breeds. From the 1960s onwards further breeds of cattle, such as the Holstein, Charolais, Limousin, Simmental, and several others were imported from continental Europe.

Different breeds are bred for different purposes. Specialist dairy breeds, such as the Holstein, are not especially good at producing meat, and cows of a beef breed such as the Aberdeen Angus have not been bred to produce large quantities of milk. But these specialist beef animals will begin to lay down subcutaneous and intramuscular fat at lower weights than the specialist dairy breeds, so that they can be 'fattened off grass'. In other words, they reach slaughter weight without a lot of concentrate feeding.

Merino sheep have been bred to produce large quantities of fine wool, but give little milk, few lambs, and poor meat, whereas a prolific Mule or Scottish Halfbred ewe, crossed with a Suffolk or Texel ram, has enough milk to raise twin lambs of high carcass quality. Sheep breeders in the US distinguish between ewe breeds such as the Rambouillet, Corriedale, Targhee, or Border Leicester, noted for their prolificacy, size, milking ability, and longevity, and ram breeds such as the Suffolk, Cheviot, or Texel, which possess good growth rate and carcass characteristics.

Modern hens bred to lay large quantities of eggs do not appear on supermarket shelves as oven-ready chickens at the end of their laying lives. They are 'spent hens', useful only for processing or pet food. Riding horses and ponies, and riding camels, are of a different shape and size from carthorses and pack camels. All of these thousands of different domesticated breeds are descended from the same original wild ancestors, but over thousands of years of selection by humans an enormous range of animals with different shapes, sizes, and attributes has emerged.

Traditional breeders worked by eye and memory. That is to say, they chose the males and females that they wanted to mate according to the characteristics they wished to produce. They expected a larger than normal bull mated to a larger than normal cow to produce big fast-growing offspring, or a high-yielding dairy cow mated to the son of another high-yielding cow to produce high-yielding daughters. Their expectations were fulfilled often enough for them to produce the traditional breeds that are still to be seen on modern farms.

By the end of the 19th century in Britain there were breed societies that kept records of the ancestry of individual animals, and shows in which breeders competed against each other to produce the best horses, cattle, sheep, and pigs, which sold for high prices. The ideal animal for the breed was described in terms of its colour and shape. In the 20th century this process was taken further by recording milk yields, which enabled breeders to select animals to mate on the basis of the performance that they wished to maximize. This was a form of performance testing, which was obviously more difficult to carry out on meat-producing animals. Carcasses don't breed! However, it is possible to weigh both animals and their food, and so measure their growth rate and feed conversion efficiency, and within the last forty years or so ultrasonic scanners have been developed which can measure the backfat level on live animals, so enabling assessment of their carcass composition.

Measurement of characteristics which are not expressed in the animal, such as the effect on milk yields of the genes transmitted by male animals, can be measured by testing progeny. For example, a bull's genetic ability to sire high-yielding daughters can be assessed by comparing the yields of his daughters with those of other bulls in the same herds, a process known as contemporary comparison.

Some of the characteristics of an animal are controlled by a single pair of genes. The black coat colour of an Aberdeen Angus, for example, is always dominant over the red coat of a Hereford, so Angus/Hereford cross cattle are always black with a white face. Most of the economically important attributes of animals, however, are controlled by many genes, so the outcome of a single breeding event is less predictable.

Three factors therefore have a big influence on the extent to which breeders can bring about genetic change in an animal population: heritability, gestation period, and the number of animals available. Heritability expresses the proportion of any difference between the individual and the population average that can be attributed to genetics rather than environment, on a scale of 0 to 1. A high score means that the offspring resembles the parent. For example, the heritability of the lactation yield is between 0.2 and 0.3, whereas that of the fat content of the milk is between 0.5 and 0.6. As a result it is easier to change the average fat content in the milk of a dairy herd than it is to increase its output.

The gestation period—the length of time the offspring spends in the womb (see Table 3)—also matters, because the faster a new generation is available to breed the more rapidly the breeder can bring about genetic change. Finally, the greater the number of animals available to the breeder, the greater the selection pressure can be. The best animal selected from one hundred is likely to be better than one selected from ten. We would therefore expect genetic improvement to be more rapid in poultry and pigs than in cattle. This is because two generations of pigs or poultry can be produced in a year, and many animals can be kept on a small area.

**Table 3 Approximate gestation periods**

|         | Average days |                          |
| ------- | ------------ | ------------------------ |
| Camel   | 406          | 12–15 months             |
| Horses  | 340          | more than 11 months      |
| Buffalo | 300–40       | 10–11 months             |
| Cattle  | 283          | 9 months 10 days         |
| Sheep   | 150          | about 5 months           |
| Goats   | 156          | about 5 months           |
| Pigs    | 116          | 3 months, 3 weeks, 3 days |

It can be as long as three years between one cattle generation and another, and only a couple of animals can be kept on each hectare.

This rapid change in pigs and poultry is indeed found in practice. In the US, between 1947 and 1988, poultry breeders halved the time and the quantity of feed needed to get a broiler chicken ready for slaughter. A 2 kg broiler can now be produced in six weeks. Over roughly the same period the growth rate of beef animals increased too, but only by about 16 per cent. Most commercial pigs and poultry are now bred by specialist breeding companies employing specialized geneticists.

Cattle breeding has not yet gone as far in specialization, but artificial insemination (AI) has helped dairy farmers, especially, to have access to high-quality genetic material. It means that rather than having to keep their own bull, they can use semen (which arrives in frozen form) from a bull with tested breeding performance. Since AI uses bulls selected under high selection pressure, genetic improvement is more rapid than is achieved using natural service. Additionally, AI allows superior males to be used on many more females. Some specialist breeders also use embryo transfer techniques, in which multiple embryos from high-quality cows are flushed out of the original cow and transplanted individually into recipient animals, thus allowing lower-performing cows to have high-quality offspring.

## Housing farm animals

In addition to feeding and breeding their animals, farmers also attempt to increase their productivity by housing them. Farm animals are quite capable of living outside all of their lives, although hens like to lay their eggs in nests, and some sows living outside will, if they can find enough suitable material, build large nests in which to give birth to their piglets. But having fed their animals, farmers are often reluctant to see part of the energy content of the ration going into keeping them warm during periods of cold, wind, and rain. This is a particularly deadly combination for young lambs born outside, even though they are perfectly capable of surviving on a clear, still, cold night. Conversely, in hot climates animals may need protection from the sun, and water buffalo need access to water in which to wallow during the heat of the day.

The sort of accommodation provided on farms is enormously variable. At one extreme there are simple buildings, little more than a roof, to keep the rain off; at the opposite extreme are controlled environment houses with artificial lighting, ventilation, and temperature control in which fattening pigs and broiler chickens may spend the whole of their lives.

In a sense this level of housing is as much for the benefit of the farmer as the animals. In an egg battery house it is easier to automate the collection of the eggs and the feeding of the birds than it is for a flock of free-range hens, thus reducing labour costs. In addition, egg laying by hens is stimulated by increasing day length, so if the birds are reliant on artificial lighting the farmer has greater control over their laying pattern. Even simple accommodation for cattle means that the farmer can decide where the farmyard manure that they produce will be used, and indeed this was one of the early arguments for housing fattening cattle. However, there is little doubt that housing, especially intensive housing for the whole of an animal's life, affects its behaviour, and this has led to arguments about animal welfare which are discussed further in Chapter 5.

Housing also has health implications. Although a farmer may house animals to protect them from the weather, the result may be to expose them to increased risks of disease. It is common, for example, to treat housed poultry with drugs to inhibit the development of various species of the *Eimeria* parasite that cause coccidiosis, an infection of the intestine. Crowded housing may also induce stress in animals, as does hunger and thirst, and a stressed animal is more vulnerable to disease.

## Farm animal health and disease

Farm animals can suffer from as wide a range of diseases as humans, with the added difficulty that they cannot tell the farmer when they are feeling unwell. One of the attributes of a good livestock manager is the ability to spot the early signs of disease, so that the animal can be isolated and/or treated, and to know the circumstances that might predispose the animal to problems.

Well-fed, high-yielding dairy cows sometimes collapse with little warning a few days after calving, a condition known as 'milk fever', which, left untreated, can be fatal. The cause is a fall in the level of blood calcium, which is easily treated, as long as the farmer knows what to do, by an injection of calcium borogluconate. This is just one example of the problems that shortages or surpluses of various minerals or vitamins can cause.

Animals are also subject to the attacks of ectoparasites (external parasites) such as flies, lice, and ticks. In the south-eastern states of the US, for example, the maggots of the screw-worm fly

(*Cochliomyia hominivorax*) burrow into the flesh of cattle. There was a successful control programme for the fly which involved releasing laboratory-bred sterilized male flies at the peak of the mating season, so that the population was reduced in a few generations and by the 1980s it had been eradicated.

Attempts to control tsetse flies (various species of the genus *Glossina*) in Africa have been less successful, and there remain large areas where cattle cannot graze for parts or all of the year. The problem is not so much the fly itself, as its ability to infect those it bites with trypanosomes, small flagellate organisms that live in the blood and cause sleeping sickness in humans and a related condition (one name for which is *nagana*) involving anaemia, fever, and slow progressive emaciation in cattle.

Similarly, ticks, of which there are numerous species, may be carriers of various diseases, such as redwater fever, East Coast fever, biliary fever, and gall sickness in African cattle, which therefore need to be sprayed weekly to control the ticks. There are also internal parasites, such as liver fluke and roundworms, bacteria, such as the clostridial bacteria that cause tetanus, and viruses, the causes of cattle plague and foot and mouth disease. All of these are potential sources of farming problems.

In the last half century or so the veterinary profession and the pharmaceutical industry have created numerous products to help farmers to combat these diseases. There are anthelmintics to control roundworms, vaccines against clostridia, and antibiotics against all kinds of bacterial infections. The question that individual farmers have to answer is whether the cost of treatment is more or less than the loss of profit from a disease, and there seems little doubt that in some circumstances they take the decision that it is worth accepting the risk of some deaths in their flocks and herds rather than bearing the cost of preventative medicines.

For some diseases, however, such as foot and mouth disease, the state often takes the view that the potential effects of disease are so great that something must be done about it, up to and including the slaughter of the affected animals. This is especially so in the case of 'zoonoses', diseases that can potentially spread from animals to people. In the last few years there have been numerous examples of these in Britain: bovine spongiform encephalopathy (BSE), Highly pathogenic avian influenza (HPAI, or bird flu), *E.coli* O157, *salmonella*, *campylobacter*, *listeria*, and bovine tuberculosis (bTB).

## Animal production systems

Taking all these considerations of feeding, breeding, housing, and disease control into account, the farmer then has to decide upon a system of production. There are almost as many different ways of keeping animals as there are different farmers, but most of them can be classified in terms of intensity, specialization, and scale. An intensive system is one which concentrates animals on a relatively small land area but uses a lot of labour, equipment, housing, and concentrated feedstuffs, whereas an extensive system uses more land and fewer of the other inputs. Thus most commercial pig and poultry units in developed countries would be described as intensive, whereas nomadic cattle herders in east Africa, or cattle ranchers in the Dakotas in the US, operate extensive systems.

Some specialized livestock farms may have only pigs or poultry, whereas on mixed farms there may be several animal and crop enterprises, as in the traditional mixed European farm, with dairy and beef cattle, sheep, a few pigs, and poultry, all fed from the grass, root crops, and cereals produced on the farm. A specialized Western European dairy farm, in contrast, would only have dairy cows, perhaps with some young stock as herd replacements, with most of the land devoted to grass, and some maize for silage.

The scale on which livestock farming takes place is almost infinitely variable, from the part-time holding with a few pigs and chickens to large units with thousands of animals on thousands of hectares. Nevertheless, most livestock farmers, no matter what the scale, intensity, or degree of specialization of their holdings, will have some of the same jobs to do.

All livestock farmers need to ensure that their animals have access to food and water every day. For some this is easy, because their animals are grazing on fields or range lands, and they may not even check them every day. Or it may simply involve pushing a button to bring automatic feeders into operation. But for others, especially in winter in temperate climates, or in the tropical dry seasons, it may take up much of the day, needing them to move large quantities of hay or silage and concentrate feeds.

If they have milking animals, they also need to ensure that they are milked every day. A cow gives milk for nine or ten months after she has had a calf (for a goat it can be much longer, and some will even give milk without having had young), and the simplest way of removing the milk from the udder is to leave it to the calf to do the job. This is the way some beef producers, with what are called suckler cows, operate, and it gives the calf a good start in life, but obviously it produces no milk for human consumption or sale. Moreover, the calf needs nothing like the milk output of a modern West European specialist dairy cow, averaging over 7,000 litres in a 305 day lactation.

For thousands of years, therefore, farmers have been restricting the calf's access to the cow, and drawing off at least some of her milk for themselves. For most of that time they have done it by hand, sitting by the side of the cow in the morning and late in the afternoon and using the actions of their hands to imitate the sucking of the calf. It was not until the 20th century that a successful milking machine was invented, but once it was available it allowed farmers to keep a larger number of cows without needing to employ many more milkers. In recent years completely automatic milking machines have been developed, which means that the cows can choose for themselves when they wish to be milked. Many of them choose to be milked more than twice a day.

Livestock farmers need to be aware of the normal behaviour of their animals. Cows reveal their readiness to be mated by their willingness to be mounted by other cows in the herd. If they are running with a bull this is mostly of interest to him, but it matters to the farmer when the cows are artificially inseminated, because getting the insemination at the right time, when the cow is 'on heat', determines whether or not she will conceive. Heat detection in sheep is much more difficult for humans, and has to be left to the ram.

Knowing when an animal is about to give birth is also important. Cows will often go away from the herd into a corner of a field, and sows kept outdoors will collect bits of straw and twigs as if they were going to build a nest. In large herds this individual observation is not always possible, and pregnant animals may be moved into calving ('farrowing' is the equivalent English term for pregnant sows) accommodation according to the length of time since they were mated. Abnormal behaviour can also be a sign that an animal is suffering from disease, and part of the farmer's job is to carry out routine injections to vaccinate animals or kill their intestinal worms, or to inspect, and when necessary trim and medicate, sheeps' feet to prevent lameness.

Cleanliness is an important part of disease control, and farmers often spend a lot of their time moving dung or cleaning housing between batches of animals so that infective organisms are not transferred from one lot to another. They will also move stock to clean pasture for the same reason, and ensure that they stay in the correct fields, which means time must be spent on the maintenance of fences and hedges.

A farmer also needs to know when animals destined for the meat market are ready for slaughter, which might simply be a matter of checking that they have reached the weight that the slaughterhouse has specified, but can also require some assessment of the subcutaneous fat cover by feeling the bones through the skin along the back.

When all those tasks have been completed, grazing livestock farmers have to ensure that they have conserved enough fodder during the growing season to last their animals though the winter or the dry season.

Although many modern animals, especially pigs and chickens, but increasingly also dairy cows, may never see a green field, their basic biology still determines the way in which farmers use them to produce food and fibre. As such, they can never forget that they are living organisms, and not simply commodities.

# Chapter 3
# **Agricultural products and trade**

As we have seen in Chapters 1 and 2, farmers spend much of their time growing crops and raising livestock, usually in excess of their family's needs. Part of their time has to be spent selling the surplus. This chapter is about what the agricultural industry as a whole produces, and about the working of the markets into which its output is sold, both local and international.

It's obvious that farmers produce food for themselves and other people, but they also produce livestock feed for farm animals (and companion animals—pets—too). These are the two biggest categories of farm product, but the list goes on: plant fibres, such as cotton or jute; animal fibres, mostly wool; fuel in various forms, from biogas to dried camel dung; pharmaceuticals, such as insulin extracted from the pancreas of pigs, and Cerebrolysin from their brains, used in the treatment of Alzheimer's disease. Where the list stops depends upon where we put the boundary around what we call farming. It would certainly include tobacco growing, but what about the cultivation of opium poppies? We perhaps think of flowers as being grown by gardeners, or, on a commercial scale, on smallholdings by people we describe as 'market gardeners', but increasingly they are grown on a field scale, as anyone who has driven round parts of the Netherlands will have seen; one might almost say on a factory scale, given the size of some of the greenhouses there. The same applies to many kinds of salad crops, such as tomatoes, peppers, and lettuces. And what about plantation crops, such as pineapples, tea, or rubber trees? Or those from small trees or shrubs, such as coffee or cocoa?

It is also important to remember that farmers don't just produce physical products. They are responsible for providing what are often classified together as 'ecosystem services': farmland is a habitat for many different kinds of wildlife, farmers may use (or allow other people to use) their land as a site for solar panels or wind-powered electricity generation, and many farming landscapes are beautiful to look at, so that farming produces a resource for the tourism industry.

The relative importance of fuel, pharmaceuticals, ecosystem services, and so on, as compared to food production, is difficult to determine, certainly on a world scale. But there is little doubt that food is the main and most important output of farming. The smaller the farm, and the less developed the agricultural economy, the more likely this is to be the case. Vegetables, salad crops, fruit, nuts, potatoes and other roots, milk, eggs, and chickens can all be grown and processed to the stage at which they are ready to be eaten without needing any more equipment or expertise beyond that normally available within the farm household.

However, as Table 4 shows, the great bulk of the world's agricultural output comes in the form of products that need more processing than most households can now manage. Moreover, since there are now more people living in towns and cities than in the countryside, the majority of people will not be able to grow their own food. Therefore, as soon as farmers produce more food than their immediate family can consume, they should be seen as suppliers of raw materials to the food industry, and the greater the scale on which they operate, the more likely this is to be the case. Even potatoes and onions, which are ready for the kitchen as soon as they leave the ground, or apples, eatable fresh from the tree, need to be stored to increase the season for which they are available, and transported to a place where the consumer can buy them.

The more industrialized the economy, the more complex that apparently simple process becomes. Apples will be sorted, graded, and perhaps have those little labels that tell you their variety stuck on them. They may then need to be stored, perhaps for several months, before being sent to a central distribution depot, and then finally to an individual shop or supermarket.

At the opposite extreme, entire industries have developed to process cereals, milk, and meat. Farmers sell their wheat to millers, who grind it into flour (and various grades of bran that can be sieved out and sold as animal feed) which they then sell to bakers. If the wheat is not of the right variety or quality to be milled for bread or biscuit flour, it will be sold to animal feed manufacturers. Milk may be consumed straight from the cow or goat, but for people in developed economies it will have been chilled on the farm, transported from the farm to the dairy, pasteurized or sterilized, and put into bottles or plastic containers. Not all of it will be needed for the liquid market, so it will be converted into butter, or yoghurt, or one of the hundreds, if not thousands, of varieties of cheese that are made.

Think of the problems that an individual farmer would encounter in trying to convert a fat bullock, weighing up to 600 kg and producing 330 kg of bone-in beef, into something that could be cooked in an oven or frying pan. Even if it could be killed effectively and humanely, there would still be the problem of dealing with the less edible parts of the carcass, of preventing the meat from going rotten before it could be eaten, and of separating the various parts of the carcass into joints that respond best to different kinds of cooking. That is why butchery has been a skilled and separate trade from farming for hundreds of years, and why people have developed a range of food preservation methods, from the traditional drying, salting, smoking, and pickling to the more recent chilling, freezing, and even radiation treatment.

Farming therefore needs to be seen as part of the food chain, with many farmers producing a relatively small range of basic products that are then transformed into the enormous range of dishes that we put on the table either in the household kitchen or by the food processing and distribution industry. The crops accounting for the greatest sown area and output are listed in Table 4, and the main animal products in Table 5.

Note that the production and yield figures given in Table 4 are fresh weights. This does not make much difference in the case of cereals and pulses, but is significant for roots and tubers. Whereas rice and wheat, in store, contain about 12 per cent water and 88 per cent dry matter, potatoes have only about 20 per cent dry matter. Taking the world average yield figures, the right-hand column in Table 4 shows the effects of these differences for some important crops. Some farmers, of course, will be able to produce more than the world average: most European farmers would not be happy unless they were producing twice the world average wheat yield, and the world record wheat yield, set in New Zealand in 2010, is 15.7 tonnes per hectare.

Table 4 shows only some of the major crops. It does not, for example, show the area of grassland, which, according to one calculation, covers 3,500 million hectares across the world, a figure which includes 35 million hectares devoted to specialist fodder crops such as alfalfa. The precise figure is a matter of definition, because there are large areas of rough grazing, savannah, etc. that are not enclosed and which farmers' animals share with a wide variety of wildlife. Table 4 concentrates on the crops which are grown for human and animal food, but there are also about 35 million hectares of fibre crops, more than 90 per cent of which are devoted to cotton.

**Table 4 The principal crops: world area, production, and average yield**

| | Area (million ha) | Production (million tonnes) | Yield (tonnes per ha) | Dry matter yield (tonnes per ha) |
|---|---|---|---|---|
| *Cereals* | | | | |
| Wheat | 215 | 670 | 3.1 | 2.7 |
| Maize | 177 | 872 | 4.9 | 4.3 |
| Rice | 163 | 720 | 4.4 | 3.9 |
| Barley | 50 | 133 | 2.7 | |
| Sorghum | 38 | 57 | 1.4 | 1.1 |
| Millet | 32 | 30 | 0.9 | 0.7 |
| Oats | 10 | 21 | 2.2 | |
| Rye | 6 | 15 | 2.6 | |
| *Sugar crops* | | | | |
| Sugar cane | 26 | 1832 | 70.2 | 7.9 |
| Sugar beet | 5 | 270 | 55.0 | |
| *Roots and tubers* | | | | |
| Cassava | 20 | 263 | 12.9 | 4.9 |
| Potatoes | 19 | 365 | 18.9 | 3.8 |
| Sweet potatoes | 8 | 103 | 12.8 | 3.8 |
| Yams | 5 | 59 | 11.7 | 3.5 |
| *Oil crops* | | | | |
| Soybeans | 105 | 242 | 2.3 | |

(*continued*)

**Table 4 Continued**

| | Area (million ha) | Production (million tonnes) | Yield (tonnes per ha) | Dry matter yield (tonnes per ha) |
|---|---|---|---|---|
| Rapeseed | 34 | 65 | 1.9 | |
| Sunflower | 25 | 37 | 1.5 | |
| Oil palm fruit | 17 | 250 | 14.5 | |
| Olives | 10 | 16 | 1.6 | |
| *Pulses* | | | | |
| Beans, dried | 29 | 24 | 0.8 | |
| Chickpeas | 12 | 11 | 0.9 | |
| Cowpeas | 11 | 6 | 0.5 | |
| Lentils | 4 | 5 | 1.1 | |
| *Fruit/vegetables* | | | | |
| Grapes/vines | 7 | 67 | 9.6 | |
| Bananas | 5 | 102 | 20.6 | 6.2 |
| Apples | 5 | 76 | 15.8 | |
| Tomatoes | 5 | 162 | 33.7 | |
| Onions | 4 | 83 | 19.7 | |
| *Beverages* | | | | |
| Coffee, green | 10 | 9 | 0.8 | |
| Cocoa, beans | 10 | 5 | 0.5 | |
| Tea | 3 | 5 | 1.5 | |

FAO.FAOSTAT. Production quantities (FAO2015) Accessed 26/7 April 2014 <http://faostat3.fao.org/browse/Q/QC/E>.

Note that area and production figures are rounded up, but yield figures are as original. For area comparisons, Great Britain is 22.7 million ha, France 55 million ha, Kenya 58 million ha, and Mexico 198 million ha.

It is important to remember, too, that some crops have multiple uses. Maize (known as corn in North America) is a good example. We are familiar with the corn cobs, sold fresh as sweetcorn, or canned as corn kernels. Steamed and crushed they make flaked maize, used as livestock feed, and with the addition of various flavourings they become cornflakes at breakfast time. They also produce starch, oil, and corn syrup. The whole crop, cobs, stalks, and all, can be harvested, chopped, and made into silage, now one of the main winter feeds for cattle in the US and Europe. And most recently, in response to higher oil prices and a legislative requirement to increase the use of renewable resources, an increasing proportion of the crop in the US has been converted into ethanol for incorporation into gasoline. By 2010 more corn was being used to make ethanol than was used to feed livestock.

There is a similar story for soya beans, which contain up to 20 per cent oil and 40 per cent protein. They are crushed to separate the oil from the remainder of the bean, which as soya bean meal is used as a livestock feed. It is the most widely used edible oil in the US (usually labelled as 'vegetable oil') and forms the basis of margarine, mayonnaise and other salad dressings, various baked products, and numerous other uses, but by 2011 nearly a quarter of all the soya bean oil produced in the US was converted into biodiesel. Similar stories could be told about other crops; farmers are not only part of the food chain, but of many other product chains too.

Livestock farmers also use a relatively narrow range of animals (see Table 2) to produce a much wider range of food and other products. Those accounting for the greatest volume of production are listed in Table 5. Comparing this with Table 2 we can see that the much greater number of chickens than any other form of livestock does not translate into greater meat production, simply because of their size.

Perhaps more surprisingly, although there are more cattle than pigs in the world at any one time, and the cattle produce two or

**Table 5  Major livestock products, world total 2012 ('000 tonnes)**

| Meat | | Milk | |
|---|---|---|---|
| Pigs | 108,507 | Cows | 626,184 |
| Chickens | 92,731 | Buffalo | 97,942 |
| Cattle | 62,737 | Goats | 18,002 |
| Sheep | 8,481 | Sheep | 10,010 |
| Turkeys | 5,634 | Camels | 3,001 |
| Goats | 5,294 | | |
| Ducks | 4,358 | *Eggs* | |
| Buffalo | 3,594 | Hens | 66,293 |
| Geese | 2,798 | | |
| Horses | 766 | *Wool* | 2,067 |
| Camels | 511 | | |

FAO.FAOSTAT. Livestock primary (FAO2015) Accessed 9 October 2014 <http://faostat3.fao.org/faostat-gateway/go/to/browse/Q/QL/E>.

three times as much meat per carcass, pigmeat (i.e. pork, bacon, ham, and all the various forms of preserved pigmeat) production is nearly twice as great as beef and veal output because pigs are ready for slaughter (at about six months) more quickly than cattle, which take at least twelve months for small cereal-fed animals, and more commonly eighteen to twenty-four months for grass-fed cattle. And of course quite a lot of the cattle population are dairy cows which would be unlikely to be slaughtered before their fifth year, and could last much longer.

China is the biggest producer of pig, chicken, sheep, and goat meat, the US of beef and turkey meat, and India and Pakistan dominate buffalo production. Camel meat is the only product in which African countries dominate the market, with Sudan, Somalia, Kenya, and Egypt among the larger producers.

Over the twenty years between 1992 and 2012 the total world output of many of these animal products increased significantly: chicken doubled, eggs nearly doubled, goat meat production increased by about 85 per cent, and pig production by about 50 per cent. Despite these increases, roughly four billion people rely on rice, maize, or wheat as their staple food. Another 500 million rely on cassava (manioc), a root crop that originated in South America but is now widely grown and eaten in sub-Saharan Africa. It is drought-resistant, so it is often grown as a 'famine crop', to be left in the ground and harvested when other staples are scarce. In some African and Asian countries such as Mali, Niger, Bangladesh, and Cambodia, cereals supply over 70 per cent of dietary energy, whereas in North America and Western Europe the comparable figure is less than 30 per cent.

Most people in these richer countries have a diet that supplements the carbohydrates supplied by cereals and potatoes with proteins and fats from meat and dairy products. But of course the animals that produce these products have to be fed, and while some of their feed comes from grass and other high-fibre plants that people cannot digest, some comes from cereals and vegetable protein-producing crops such as soya beans. Over 70 per cent of the cereals grown in Canada in 2005, and 40 per cent of the wheat grown in Britain, went to feed animals. On a world scale, it is estimated that about a third of all cereals are fed to animals. Soya bean production in Argentina and Brazil, taken together, nearly quadrupled between 1993 and 2013, as their cultivation expanded into areas previously used for grazing or covered in forest, and about 80 per cent of the world's soya bean harvest is used for animal feed.

Does this mean that Argentina and Brazil have dramatically increased the number of animals that they produce, and are consequently consuming much more meat and more dairy products? Not necessarily, because, as we saw earlier, it is important to see agriculture as a producer of raw materials for the food industry, or as the first link in the food chain. Thus any farm product might be consumed on the farm, within a local market area, perhaps in the country in which it was produced, or in some other part of the world. In order to understand why this happens we need to know more about how agricultural markets work, and that is what we consider in the next section of this chapter.

## The market for farm products

The county of Kent in south-eastern England is traditionally noted for the extent and quality of its fruit farms. In the autumn of 2014 apple growers there were expecting a bumper harvest. The previous winter had been cold, keeping pests at bay, there was enough rain in August to help the fruit to grow, and a beautiful sunny September enabled picking to begin early. It was much the same in the rest of Europe. According to newspaper reports, one farmer couldn't remember a better growing year, and another spoke of getting 'quite a buzz, seeing a fantastic crop'. But they both felt that their apples would be worth little, and one was preparing to make a loss while the other felt that he might just cover his costs. This seems crazy, but it's not a new problem. The Porter in Shakespeare's *Macbeth* imagines himself opening the gates of hell to 'a farmer that hang'd himself on th' expectation of plenty'. So how can a wonderful crop lead to a financial loss? If we look at the way in which agricultural markets work we should be able to explain it.

As with any other market, agricultural markets are affected by changes in demand and supply. If demand increases but the supply remains unchanged, prices are likely to rise; if supplies increase but demand remains constant they are likely to fall. We therefore need to know what affects the demand for and supply of farm products.

The demand for a product is influenced by its price, the incomes and tastes of the potential consumers, and the number of those consumers. When prices are high consumers think hard before buying a product, and when they are low they consider buying more of it. That's the basic idea, but of course it also depends upon what the product is.

For staple foods, such as rice, maize meal, pasta, bread, or potatoes, the demand will not change much. When prices are high people still need the basics of their diet, so they will pay what they have to. And when prices are low they are not likely to consume vastly increased quantities of their staple foods, but instead use the money that they have saved on some other product. Conversely, the market for luxury foods will be much more affected by the price level, because they are not a vital part of the diet, so a small price increase might cause a big decrease in the quantity consumed, and vice versa. The possibility of substitution has an impact here too. We can obtain protein from the more expensive cuts of meat, which are tender and taste good, but we can also get it from cheaper cuts. Similarly, high beef prices might increase the demand for pork, as long as consumers see roast pork as an acceptable substitute for roast beef. However, some consumers may have cultural or religious objections to eating pigmeat in any form, so their preferences would not affect the pork market.

This is just one of many examples of the way in which consumer tastes affect demand. Governments in many industrialized countries have attempted to influence consumer food preferences for health reasons, encouraging, for example, the consumption of less red meat and more fruit and vegetables. Whether these campaigns have been as successful as commercial advertising, which is the other main influence on consumer tastes, is a matter of controversy (which is further discussed in the Very Short Introductions on *Food* and *Nutrition*).

Income levels affect the demand for food in all sorts of ways. A study carried out for the United States Department of Agriculture in the early 21st century compared the behaviour of consumers in low-income countries, with incomes less than 15 per cent of those in the US, and high-income countries, with incomes of 50 per cent or more of the US level. Some of the results of the study are given in Table 6. They show that people in richer countries spend a smaller proportion of their incomes on food than those in poorer countries, and that their food consumption behaviour is less responsive to income changes. As people get richer they spend

**Table 6  Income effects on the demand for food**

|  | Low-income countries | High-income countries |
|---|---|---|
| Percentage of budget spent on food | 47 | 13 |
| Increase in demand for food when income increases by 1% | 0.73% | 0.29% |
| Percentage of food budget spent on cereals | 28 | 16 |
| Increase in cereal consumption when income increases by 1% | 0.56% | 0.19% |
| Percentage of food budget spent on meat | 18 | 25 |
| Increase in meat consumption when income increases by 1% | 0.82% | 0.33% |
| Percentage of food budget spent on dairy products | 9 | 14 |
| Increase in consumption of dairy products when income increases by 1% | 0.93% | 0.35% |

A. Regmi, M. S. Deepak, J. L. Seale Jr, and J. Bernstein, *Cross Country Analysis of Food Consumption Patterns*, Economic Research Service, US Department of Agriculture., tables B-1 and B-2, available at <http://www.ers.usda.gov/media/293593/wrs011d_1_.pdf> (accessed 20 October 2014).

more on foods containing proteins and fats, like meat and dairy products, but consumption of staple foods such as cereals increases less rapidly. The study also found that, perhaps not surprisingly, the poor are more responsive to food price changes than the rich.

These data are bad news for farmers in rich countries, because they imply that the demand for food will not increase much unless the population increases, and population increases in rich countries tend to be lower than in poor countries. Conversely, they are bad news for consumers in poor countries, because they imply that small increases in food prices or small decreases in incomes will have big effects on the amount of food that they can buy.

Does a slow demand increase matter to rich-country farmers? It does if they increase supplies more than demand, because the prices they receive will then fall. In theory, this should be a self-regulating system, because lower prices should result in less being supplied to the market, as all farmers reduce their output, or some go out of business altogether, and supplies thus fall to the level at which they equate to the quantity demanded.

In practice it's more complicated. Product prices are not the only influence on farm output. Production costs can change: pig and poultry profitability, for example, is very sensitive to changes in the price of feedingstuffs. Farmers can adopt new varieties of crops, or breeds of livestock, which produce more without needing more inputs, or any of the other myriad technical changes that increase output or reduce costs. Government policies can change, so that farmers are encouraged to produce more, as they were after World War II and for many years after in Europe, or less, as they were when this policy succeeded to the point of producing surpluses.

It is also important to remember that agricultural production cannot be turned on and off at the flick of a switch. If farmers decide that they want to produce more milk it could take them more than two years to raise heifer calves to the point at which they are ready to be mated, become pregnant, deliver a calf, and take their place in the dairy herd. Tree crops may also take years to come into production. Overall, therefore, with some exceptions, output of many farm products tends to respond only slowly to changes in market conditions.

The reason why apple growers in Kent might have mixed feelings about a bumper harvest should now be clear. The demand for apples in the UK is falling as the range of fruit sold in supermarkets increases, and falling prices will not have much impact on the quantity of apples that consumers buy. In addition, UK farmers have to compete with suppliers in other countries who also sell their apples into the UK market, and that takes us on to another important issue for farmers: the question of international trade.

## International trade in farm products

As we saw earlier, some farmers simply produce what they and their immediate families want to eat, and sell none of their output. They are subsistence farmers, and there are not many of them. Most farmers probably sell some of their produce, simply because they need things that they cannot produce for themselves, from salt and iron to fertilizers and pesticides. Farmers have sold food and fibres to non-farmers almost since the beginning of farming. Eventually this led to regional specialization, as farmers on good arable soils realized that they could make more money by specializing in crop production, and those on poor upland farms realized that they might do better by exchanging their livestock for the crops they needed. The logical end result of this process is that countries specializing in agriculture should export farm products in exchange for the manufactured goods produced in industrialized countries.

This was the pattern that had emerged by the end of the 19th century. The bulk of international trade then was in temperate products, sent from countries that had been settled by European emigrants, such as the US, Canada, Argentina, Australia, and New Zealand, back to Europe, or from Eastern Europe and Russia to Western Europe. Tropical products, such as tea, coffee, cocoa, and oilseeds, accounted for the minor part of the total.

**Table 7 The top ten importers and exporters of agricultural products in 2012**

| Exporters | % of world exports | Importers | % of world imports |
|---|---|---|---|
| EU 27 | 37.0 | EU 27 | 35.7 |
| (EU exports to non-EU countries) | (9.8) | (EU imports from non-EU countries) | (9.9) |
| US | 10.4 | China | 9.0 |
| Brazil | 5.2 | US | 8.1 |
| China | 4.0 | Japan | 5.4 |
| Canada | 3.8 | Russia | 2.4 |
| Indonesia | 2.7 | Canada | 2.2 |
| Argentina | 2.6 | Republic of Korea | 1.9 |
| India | 2.6 | Saudi Arabia | 1.7 |
| Thailand | 2.5 | Mexico | 1.6 |
| Australia | 2.3 | India | 1.5 |
| % of world total (by value) exported by the top 10 | 73.1 | % of world total (by value) imported by the top 10 | 69.5 |

World Trade Organization, *International Trade Statistics 2013, part II, Merchandise Trade*, © World Trade Organization, 2013, <http://www.wto.org/english/res_e/statis_e/its2013_e/its13_merch_trade_product_e.pdf, table II.15> (accessed October 2014).

Agricultural products and trade

Perhaps surprisingly, there are still echoes of this in the present pattern of world agricultural trade. It continues to be dominated by industrialized countries, as Table 7 shows, although some newly industrializing countries have now joined the list of major importers. Europe is still the major importer, and the US, Canada, Argentina, and Australia remain in the top ten exporting countries. The EU, the US, and China appear as both importers and exporters simply because they are major importers of some commodities and major exporters of others. The US, for example, is one of the top five exporters of wheat, rice, maize, and beef, and among the top five importers of sugar, coffee, cocoa, and tea. There is considerable trade among the EU countries, which accounts for the EU's position at the top of both lists, and traders in some EU countries act as both importers and exporters: Germany, for example, is a major importer and exporter of butter and cheese.

**Table 8 World exports as a percentage of world production in 2011–12**

| | | | |
|---|---|---|---|
| Wheat | 22 | Pigmeat | 13 |
| Maize | 13 | Chicken | 13 |
| Rice | 5 | Beef | 16 |
| Sugar | 22 | Sheepmeat | 10 |
| Potatoes | 3 | | |
| Bananas | 18 | Milk (equivalent) | 14 |
| Soyabeans | 38 | | |
| Coffee | 84 | Eggs | 3 |
| Cocoa | 64 | | |
| Tea | 40 | | |

FAO.FAOSTAT. Crop and livestock products. Exports of selected countries (FAO 2015) Accessed 10 October 2014 <http://faostat3.fao.org/browse/T/TP/E>.

Although the bulk of the tropical beverages—tea, coffee, and cocoa—are produced for export, and a significant proportion of the soya beans are exported, trade in foods generally represents only a small proportion, often less than 20 per cent, of world production, as Table 8 demonstrates. When harvests are poor, domestic requirements tend to be satisfied first, so less is available for export, whereas much of the surplus from good harvests will be sent to export markets. Thus the supplies available to the world market tend to fluctuate more than world output, and world market prices are correspondingly variable.

For example, consider the case of a hypothetical country exporting the world average percentage of rice from a harvest of 100 million tonnes. In a normal year it will supply 5 million tonnes to the world market. In a good year when production increases by, say, 5 per cent, it will have 10 million tonnes available for sale, and in a poor year, with output 5 per cent down, none at all for export if its domestic requirements are to be satisfied.

Governments are understandably concerned about these questions of international trade and domestic demand and supply. Even in industrialized countries with varied and plentiful supplies, food prices can be a sensitive political issue; in poor countries the supply/demand balance can be a matter of life and death. Thus most countries have some kind of government involvement in agriculture, which may range from statistical monitoring of inputs and outputs, or provision of advice to farmers, to comprehensive control of production, prices, and farm incomes. Even governments in those industrialized countries that favour the operation of market forces, such as those of the EU or the US, take measures to support farm incomes and affect the trade in farm products.

At an international level, the World Trade Organization monitors agricultural trade and through periodic intergovernmental meetings attempts to reduce barriers to the free flow of trade. The

Food and Agriculture Organization of the United Nations collects a comprehensive range of agricultural statistics and produces numerous reports on the state of world agriculture and what can be done to address its problems, especially in developing countries.

The purpose of this chapter has been to show that the output of the agricultural industry, with all its variety, is simply a part of the complex of local, regional, national, and international links that have emerged to try to ensure that people have something to eat every day. These linkages also extend to the other side of the agricultural industry, to provide the inputs that farmers need, and they form the subject of Chapter 4.

# Chapter 4
# Inputs into agriculture

The story is told of an old farmer who took over some land that had not been farmed for years. It was overgrown with weeds, the gates were hanging off their hinges, the hedges were either overgrown or full of gaps, the drains and ditches were blocked so that the lower fields were boggy. After several years of hard work and considerable investment the land was transformed. There were flourishing crops growing in well-drained fields with not a weed to be seen. The pastures, now surrounded by trimmed stockproof hedges and new gates, fed handsome cattle and sheep in great quantities. One day, as the farmer was leaning on the gate admiring it all, the parson came by. 'You and the Good Lord have made a wonderful farm here, John', said the parson. 'Perhaps so', John replied, 'but you should have seen it when he had it to himself'.

Left to itself, land will produce food for people, as Palaeolithic hunter-gatherers knew, but their population densities remained low. To feed the present human population, farmers use the land and add to it their labour, their animals, buildings, tools, and machinery, a variety of products bought in from other industries, from fertilizers and pesticides to computers and mobile phones, and professional services from vets, advisers, and scientists. This chapter is about those inputs. But first, the land.

## Land

It is possible to produce food without land. Crops can grow in a nutrient solution, a technique called hydroponics. Increasingly salad crops are grown in greenhouses, in which large quantities of produce come from relatively small areas of land. Intensive livestock, such as pigs and poultry, are also housed on relatively small areas of land, although of course much greater areas of land are needed to produce their feed. But for most farmers, land, 'the original and indestructible powers of the soil' in the British economist David Ricardo's words, is a basic necessity.

The world's land area is just over 13 billion hectares, of which 37.6 per cent, or nearly 5 billion hectares, is agricultural land. The rest is approximately equally split between forest land and other uses, which range from urban land to steep uncultivable mountains and land covered in ice. Of the agricultural land, 28.3 per cent is arable and 3.1 per cent in permanent crops, such as coffee or apple trees. The rest is meadows and pastures, of varying kinds and qualities.

The great bulk of the world's food comes from its agricultural land, which is far from evenly distributed. Only 44 per cent of the land in the US, one of the world's great agricultural exporters, is in agriculture, compared with over 70 per cent in Bangladesh but only 12.6 per cent in Japan. There are similar variations in the amount of cropland per head of population, from 2 hectares in Australia to 0.03 hectares in Japan. The world average in 2010 was 0.22 hectares per person, which is half of what it was in 1960. This means that food output per hectare must have grown over the last fifty years, which indeed it has, by between 2 and 4 per cent per year. At the same time the amount of cultivated land has also increased, by about 1 per cent per year, but the potentially cultivated land which is not yet farmed is very unevenly distributed. According to the Food and Agriculture Organization

(FAO) of the United Nations, there is no 'spare' land in southern or western Asia or North Africa, and 90 per cent of the accessible land is in seven countries: Congo, Angola, Sudan, Brazil, Argentina, Colombia, and Bolivia. Since its further accessibility would have environmental and social implications, we might question whether there is any spare land at all.

It should be clear from Chapter 1 that not all land is equally productive. Variations in soil, altitude, aspect (whether a slope is north- or south-facing—south-facing slopes get more solar radiation in the northern hemisphere and vice versa), and climate, especially as regards temperature and rainfall, will all affect productivity.

For example, in global terms, Tanzania is well-provided with agricultural land, with 0.77 hectare per person, but much of this land is dry savannah, suitable only for extensive grazing. Japan, in contrast, as we have seen, has very little agricultural land, but much of it is intensively cultivated to produce rice and vegetables. While Tanzania has more than twenty times as much land per person as Japan, the value added per Japanese hectare is sixty-seven times the value added on a Tanzanian hectare. Hence the idea of grading land according to its flexibility: the best land can produce more or less anything, and as land quality decreases there are increasing limits on what it can produce.

Water is often the limiting factor. In some cases the land is too wet, or wet at the wrong time of year, and needs to be drained. More commonly it is too dry, and requires irrigation. Some crops, such as certain types of rice, are entirely dependent upon irrigation, and crop production in drier parts of the world can also be reliant on irrigation.

There are well-known examples in the Central Valley of California and in the High Plains, that drought-prone area to the east of the

Rocky Mountains, part of which is underlain by the Ogallala aquifer, which extends from South Dakota to north-west Texas. Its use illustrates both the potential and pitfalls of irrigation. The development of high-volume pumps and centre pivot irrigation systems enabled the area of land irrigated from the aquifer to increase from about 800,000 hectares to over five million hectares between 1950 and 1980, and the resulting green crop circles are still clearly visible on satellite photographs of the region (see also Figure 4). By the late 1970s, what had been a drought-stricken dustbowl was producing corn, cotton, wheat, and sorghum, but by then it was clear that the Ogallala aquifer was being emptied far faster than it was being filled, and, as also happened in parts of California, the water table was dropping.

There are other examples of unsustainable extraction on the plains of northern China and the Indo-Gangetic plain in India. This has led to the concept of the 'water footprint', analogous to the carbon footprint: the idea that different crop and animal products require very different quantities of water in order to provide the nutrients that we derive from them. Most (78 per cent according to one calculation) of the water used in global crop production is rainwater, which is called 'green water' in these calculations. 'Blue water' is the surface and groundwater, which either evaporates, flows to the sea, or is used in agriculture. It provides about 12 per cent of the water used in crop production. The third category is 'grey water', which is the volume of water needed to assimilate any pollutants, and it accounts for the remaining 10 per cent of the water use.

Farm animals too use water, for drinking, obviously, but a lot is also used in cleaning their housing and providing the food they eat. Overall, the agricultural sector accounts for about 85 per cent of global blue water consumption. Table 9 shows the considerable variation in water footprint of various sources of energy, proteins, and fats.

**Table 9 Water footprint per unit of nutrition**

| Energy (litres per kcal) | Protein (litres per gramme) | Fat (litres per gramme) |
|---|---|---|
| Beef 10.19 | Nuts 139 | Starchy roots 226 |
| Sheep/goat meat 4.25 | Beef 112 | Pulses 180 |
| Chicken meat 3.00 | Pigmeat 57 | Beef 153 |
| Sugar crops 0.69 | Eggs 29 | Milk 33 |
| Cereals 0.51 | Pulses 19 | Pigmeat 23 |
| Starchy roots 0.47 | Oilcrops 16 | Oilcrops 11 |

Data from M. M. Mekonnen and A. Y. Hoekstra, 'A Global Assessment of the Water Footprint of Farm Animal Products', *Ecosystems*, 15, 2012, pp. 401–15.

# Labour

For most of human history, the majority of people lived outside towns and cities. But since 2008 the world's urban population has been bigger than the rural population, and it is predicted that urban families will account for the majority of future population growth. This obviously has implications for agriculture, because the majority of agricultural workers live in the countryside, although it must be remembered that a strict distinction between agricultural and non-agricultural workers is more of a first-world idea than something that would be recognized in developing countries, where many city dwellers—perhaps over 40 per cent of the population of Ouagadougou, the capital of Burkina Faso, for example—are still involved in the production of crops and livestock.

Taking the world as a whole, over a quarter of the population (i.e. the economically active population and their non-working dependants) is employed in agriculture, but within this total there are great variations from one country to another. Whereas the figure for Africa as a whole was 49.1 per cent in 2010, and for India it was 48.4 per cent, in Latin America it was 15.8 per cent,

6. **Farmers in Somoa village, upper west Ghana, sell their farm products, including millet (on the right of the picture), at market.**

and in the US, Canada, and the UK it was only about 1.8 per cent of the population. However, quoting these international statistics raises a question that is not as easy to answer as it might seem at first sight: who should be included in the category of agricultural worker, and what counts as a farm?

Throughout this book we have referred to the people carrying out agricultural operations as 'farmers', but this is the point at which we should recognize that we have used the term to cover the activities of a wide range of people. It is certainly true that most farms in most countries are owned by an individual or group of individuals and use mostly household labour (see Figure 6). In other words, they are family farms.

Having said that, there are other people who are involved in agriculture without being members of family farming households. Most obviously, there are those who work on a farm without having a share in its ownership. They are variously called farm workers or agricultural labourers and they work for wages. Conversely, there are those who own farm land but do not work it themselves. They are landowners, and they lease their land to tenant farmers in return for a rent. Sometimes governments, or religious institutions, or corporations act as landlords. Alternatively, corporations (sometimes referred to as 'agribusinesses') or co-operatives may employ managers and workers to farm the land rather than leasing it to family farmers. It would be quite possible to find farms on which a family owned some of the land, rented some more, and also had rights to use other land in common with other farmers, for grazing, for example.

Over time, these different roles have often acquired labels with implied status. Thus small family farmers are often called peasants. In some countries the term has connotations of taking pride in independence and self-sufficiency; in others it may imply a low status and lack of sophistication. Landowners, who do not work the land, are sometimes called landlord, a word which by itself seems to imply high status. Conversely, the implicit low status ascribed to landless agricultural labourers covers people who may be poorly educated, low-paid migrant workers, or, in some developed countries, highly trained and well-paid agricultural graduates.

As we might expect, if there are many kinds of farmers, there are also many definitions of what constitutes a farm. At first sight it seems pretty obvious that a farm is an area of land, with its associated buildings and machinery, on which people grow crops and raise livestock. But which people, how much land, and what sort of crops or livestock?

Countries vary in what they consider as the minimum size for a farm when responding to the World Census of Agriculture, which is carried out by every decade by the UN's Food and Agriculture

Organization. India has no minimum, whereas Bangladesh only included holdings of greater than 0.2 hectares, and China included farms of only 0.07 hectares. Germany has a minimum farm size of 2 hectares. In Russia, 98 per cent of the holdings are owned by private individuals, but they only cover 2 per cent of the farmland area; the other 98 per cent of the farmland is owned by institutions, entrepreneurs, and non-profit citizen associations.

Thus any figure for the number of farms in the world cannot be exact, because it depends upon definition, but the latest attempt to estimate it concluded that there are now more than 570 million farms in the world, and that the number is increasing as a result of growing numbers in low- and middle-income countries. More than half of all farms are in China and India; only 4 per cent are in high-income countries. The bulk of them—more than 500 million—are family farms, in the sense that the family owns them and does most of the work on them, and most of them are small: 94 per cent of all the farms in the world have less than 5 hectares of land.

However, looking at farm sizes through world averages hides a lot of geographical variation. In East and South Asia, and in sub-Saharan Africa, most of the farms are of less than 5 hectares. They account for at least half of the agricultural area, whereas in Latin America there are fewer very small farms and most of the land is in a few of the larger farms. In high-income countries, although many small farms remain, the bulk of the land is worked by the big farms. In the US, for example, 80 per cent of the output, in value terms, is produced by the biggest 10 per cent of farms. Table 10 gives some examples from individual countries.

Table 10 illustrates one of the difficulties in discussing farm size: if you ask farmers about the size of their farms they will almost always answer in terms of hectares or acres or whatever the local measure of land area is, or perhaps in terms of the number of livestock they have, but the question is often really about the size

**Table 10 The distribution of farms and land in the late 20th century**

|  | Ethiopia | Ecuador | UK | US |
|---|---|---|---|---|
| Number of holdings ('000) | 10,758 | 843 | 233 | 2,129 |
| Total agricultural area ('000 ha) | 11,047 | 12,356 | 16,528 | 379,712 |
| Holding size (ha) trend | Falling 1.4 → 1.0 | Falling 15.3 → 14.7 | Rising 55 → 71 | Rising 158 → 178 |
| % of holdings of less than 5 ha | 99 | 64 | 23 | 11 |
| % of land in holdings of less than 5 ha | 93 | 6 | 1 | 0.14 |
| % of holdings of 5–100 ha | 1 | 35 | 60 | 38 |
| % of land in 5–100 ha holdings | 7 | 65 | 30 | 12 |
| % of holdings with more than 100 ha | 0 | 1 | 17 | 27 |
| % of land in holdings with more than 100 ha | 0 | 29 | 69 | 88 |
| Average number of hired permanent workers per farm | 0 | 0.3 | 0.6 | 1.4 |

Data from *Food and Agriculture Organization of the United Nations*, 2014, S. K. Lowder, J. Skoet, and S. Singh, 'What Do We Really Know about the Number and Distribution of Farms and Family Farms Worldwide?', background paper for *The State of Food and Agriculture 2014*, ESA Working Paper No. 14–02, Rome: FAO <http://www.fao.org/docrep/019/i3729e/i3729e.pdf>. Reproduced with permission.

of the farm business, and whether or not it will support the farm family. A small farm on very fertile soil with intensive cropping may produce far more than a much bigger farm, in terms of land area, on poor land suitable only for grazing a few cattle.

However, international statistical organizations collect farm size data in area terms only, so it is those that we have to use. And clearly the fact that 29 per cent of the agricultural land in Ecuador is in the hands of the 1 per cent of holdings with more than 100 hectares tells us that a significant proportion of the country's agricultural output is likely to come from those farms. Similarly, in the UK, the vast majority of the farms have less than 100 hectares of land, but, as in the US, the bulk of the output comes from those with more than 100 hectares.

This often has important political implications. An agricultural policy designed to help big farms may make life difficult for small farmers. However, each of the smaller farmers has a vote, both in government elections and in elections for office-holders in farmers' pressure groups, so their concerns may achieve more political prominence than those of the bigger farmers, despite the fact that the latter may have more spare time and money to devote to political activities. On the other hand, even the biggest farms have less market power than food processors and retailers in developed countries, so they have to accept the prices that they are offered just as the smaller farmers do.

The other obvious point illustrated by Table 10 is that small farms are more likely to rely on family labour and less likely to hire labour than big farms. As the table shows, while farm sizes are increasing in the UK and the US, and in many other developed countries, in developing countries they are often decreasing as populations increase. This means that families may have too little land to use their available labour efficiently. If there are local employment opportunities in the non-farm rural economy that may be a positive feature, but otherwise it may result in permanent or temporary migration to urban areas in search of work.

Examined from the viewpoint of international statistics, and perhaps even from the perspective of national agricultural policymakers, the relationship between land and the labour that works it is complex enough, but it is worth remembering that we are talking about families that don't just work the land—they usually live and work on *their* land, a particular bit of land, which they may have worked for a long time, and which their parents and grandparents may have worked before them. Thus they have a connection with it that may go beyond the purely economic.

European farmers, asked about their business objectives, may talk about profit maximization or long-term growth, but many of them will also say that their principal purpose is to hand on a viable farming business to the next generation, just as one was given to them. Exactly the same view was articulated by a representative of Maasai pastoralists in Tanzania when the government reportedly wished to turn their land into a game-hunting reserve. They were offered financial compensation, but 'you cannot compare that with land. It's inherited. Their mothers and grandmothers are buried in that land. There's nothing you can compare with it.' Whether this will continue as family farms grow bigger, and more land in high-income countries is farmed by farming corporations, remains to be seen.

Increasingly, farmers and farm workers are assisted by numerous other specialists. Originally, perhaps, farmers made all they needed from the resources of the farm, with assistance from the blacksmith in making ploughshares, and from the miller in processing their grain. Modern farmers may have their fields ploughed and sown and reaped and mown by specialist contractors operating big machines. When their own machinery breaks down they can call for a mechanic from the local machinery dealer. Their cattle may be artificially inseminated by a travelling specialist who visits the farm, and their fat cattle, sheep, and pigs will be driven to the slaughterhouse by a lorry driver from a livestock haulage firm. An agronomist may tell them which pesticide to use on their crops, and consultants, advisers, and bank managers may discuss their business plans. New technology may be developed for them by agricultural scientists working in universities and research institutes or for commercial firms. They may be made aware of it when they are trained in agricultural colleges and universities, and subsequently by agricultural journalists and broadcasters, or by the sales staff of firms in the agrochemical and farm machinery industries.

After leaving the farm, their crops and livestock will be passed on to the food processing and retailing industries, so although the numbers directly involved in agriculture in developed countries may be a tiny proportion of the labour force, many more people are involved in the food chain as a whole.

## Other inputs

When people first began to cultivate plants to eat, as opposed to simply collecting what they found growing wild, they would have needed some kind of digging implement to make a seedbed, sickles to cut their cereals, and baskets in which to collect them. But much of the work, and the power behind it, would have come from human hands and muscles. Over time, the proportion of the work done directly by the human hand has diminished, and the importance of the other things that farmers use has increased.

The list is almost endless, from draught animals and breeding flocks and herds, to fruit trees, buildings, fences, hedges, ditches, drains, tools, machinery, fertilizers, and pesticides. They are the things that farmers use in the course of producing crops, animals, and animal products. Most of these were, at one time, produced on the farm. A harrow to rake over a seedbed could be as simple as a bush cut from a hedge, pulled by an ox or a cow. The cows in the dairy herd and the sows in the pig herd were bred on the farm. The wood to make fences or the framework of buildings could be cut from nearby woodland or the farm's hedges. The hedges themselves started as young cuttings planted by the farm staff, who also dug the ditches beside them. The only major fertilizer was the manure produced by the farm's animals, although in some places farmers might be fortunate enough to live within reach of some other nutrient source, such as seaweed. Weeds were controlled by hoeing. And in some developing countries this continues. But in developed countries, modern equipment has largely taken over. Perhaps the most iconic item is the tractor, which has almost completely replaced draught cattle and horses in developed countries.

Different kinds of farm demand different kinds of input. Land and labour will account for a relatively greater proportion of the total costs on an extensive grazing farm on which the farm workers spend most of their time looking after the animals and the land receives little fertilizer, whereas on an intensive highly mechanized arable farm the machinery, pesticides, and fertilizers will account for more of the expenditure. Table 11 gives some idea of the enormous variability between different countries in the use of some of those inputs that are more easily quantified. It shows that richer countries tend to use more purchased inputs than poorer countries, which is what we would expect, but also that input use also varies with the intensity with which the land is farmed.

These huge variations in machinery, pesticide, and fertilizer use (and in other inputs, less easily quantified, such as buildings) mean that the carbon footprint of agriculture varies from one country to another, just as it varies from one crop or livestock product to another, or one food consumer to another.

There are three principal greenhouse gases: carbon dioxide ($CO_2$), methane ($CH_4$), which has twenty-three times the global warming potential of $CO_2$, and nitrous oxide ($N_2O$), with 296 times the global warming potential of $CO_2$. Carbon dioxide is produced by the internal combustion engines powering farm machinery, in the manufacture of fertilizers, especially nitrogen fertilizers, and when land is cleared from forest or ploughed. Ruminant animals emit methane as one of the by-products of their digestive processes,

**Table 11 Purchased inputs in agriculture in various countries, 2000–10**

|  | Tractors per thousand ha of arable land | Tractors per thousand workers in agriculture | Pesticide use (kg per ha) | Fertilizer use (total of N, P, and K, kg per ha) |
|---|---|---|---|---|
| Mali | 0.2 | 0.1 | 0 | 7.4 |
| Tanzania | 2.1 | 0.8 | No data | 7.5 |
| Ecuador | 4.0 | 4.6 | 3.7 | 88.2 |
| Brazil | 12.9 | 28.4 | No data | 111.7 |
| India | 13.2 | 3.7 | 0.2 | 156.6 |
| US | 27.0 | 707.1 | 2.2 | 107.7 |
| France | 64.1 | 603.6 | 2.9 | 140.2 |
| UK | 75.6 | 976.4 | 3.0 | 238.2 |
| Japan | 472.0 | 420.6 | 13.1 | 271.1 |

FAO, *FAO Statistical Yearbook 2013: World Food and Agriculture*, Rome: FAO, 2013. Reproduced with permission.

and nitrous oxide is produced from the breakdown of fertilizers and animal manure. Thus highly mechanized agricultural industries using lots of fertilizer produce more greenhouse gases than less developed agricultures, but even low-intensity cattle grazing will have some impact on global warming through the production of methane.

There are widely varying estimates of agriculture's contribution to greenhouse gases, ranging from 17 to 32 per cent of the total. A Food and Agriculture Organization study found that the world's livestock alone accounted for 9 per cent of carbon dioxide, between 35 and 40 per cent of methane, and 65 per cent of the nitrous oxide in global anthropogenic emissions. Different foods similarly have widely varying carbon footprints, as Table 12 shows,

**Table 12  UK greenhouse gas emissions (kg of CO$_2$ equivalent per kg of product)**

| Beef | 68.8 | Coffee | 10.1 |
|------|------|--------|------|
| Mutton | 64.2 | Rice | 3.9 |
| Animal fats | 40.1 | Sunflower oil | 3.3 |
| Pigmeat | 7.9 | Wheat | 1.0 |
| Poultrymeat | 5.4 | Beans | 0.8 |
| Eggs | 4.9 | Potatoes | 0.4 |
| Milk | 1.8 | Sugar | 0.1 |

Source: Data from P. Scarborough, P. N. Appleby, A. Mizdrak, A. D. M. Briggs, R. C. Travis, K. E. Bradbury, and T. J. Key, 'Dietary Greenhouse Gas Emissions of Meat-Eaters, Fish-Eaters, Vegetarians and Vegans in the UK', *Climate Change* 125, 2014, pp. 179–92.

Note that much lower figures for Norway are shown in D. Blandford, I. Gaasland, and E. Vårdal, 'Extensification versus Intensification in Reducing Greenhouse Gas Emissions in Agriculture: Insights from Norway', *Eurochoices* 12(3), 2013, pp. 4–8.

with animal products and imported commodities having higher figures than plant products.

Thus the carbon footprints of individual consumers vary with their diets. A UK study found that those eating more than 100 grammes of meat per day (about two chipolata sausages plus two slices of roast beef) would be associated with emissions of 2,624 kg of CO$_2$ equivalent per year, whereas the figure for vegetarians was only 1,391 kg. (For comparison, the carbon footprint of a ten-year-old small family car covering 6,000 miles per year was 2,440 kg.) Will both farmers and food consumers come under pressure to reduce these carbon emissions in the future?

It is important to remember that farmers do not just use *things* to help them to cultivate the land and care for their animals. Tradition, experience, and knowledge are important too. Knowing when a crop or an animal is doing well, or spotting problems in time to do something about them, are an important part of a farmer's everyday job. Many farmers acquire ideas about the importance of good livestock care or 'doing right by the land' as part of their upbringing or training in farming. An economist would say that it was part of the social and intellectual capital of the industry. As with the other kinds of capital, such as machinery and buildings, it can be increased by investment, in this case in education and training or research and development. Many countries now have at least one, and in some cases a network, of agricultural colleges and university departments of agriculture, offering both full-time courses and other training opportunities. Many of them also carry out research, or are linked to other research institutions.

Whereas roughly fifty years ago most of the world's agricultural research was carried out in developed countries, the present pattern is that an increasing proportion of research spending is found in the newly industrializing countries. The combined agricultural R&D spending of China, India, and Brazil in 2008 was $US7.6 billion, compared with a total of $US9.2 billion in the US, Japan, and France. It is probably too soon to say how this will affect farming in those countries, and in developing countries. The experience of the agricultural industries in North America and Western Europe suggests that it takes time for effective knowledge networks to develop and transmit new technologies.

## Farming systems

If we now bring together the information about the outputs of farming from Chapter 3 with the data on inputs that we have compiled in this chapter, we should be able to see how differences can combine to produce different farming systems.

This is something that has fascinated some academic experts on agriculture for many years, and they have produced a variety of more or less complex analyses, coupled with the lessons that might be learned from them for purposes of farm management and agricultural policy making. The basic ideas, however, are fairly simple, and we have implicitly used them already. Each farm differs according to where it can be placed on several continuously varying scales: intensive/extensive, simple/advanced technology, large/small size, mixed/specialized, tropical/temperate, close to/ far from markets, and so on. Chapters 5 and 6 will look at the ways in which these differences affect farming as it is presently carried out, and how farmers might deal with the problems they have to face in the future.

# Chapter 5
# Modern and traditional farming

Your ideas about the difference between modern and traditional farming probably depend on where you live, and when. People old enough to remember back to the 1950s who live in paddy rice-producing areas might think of traditional farming as involving tall crops cultivated with the help of animals using only organic manures. Where yams are a staple part of the diet they might think of crops grown on small hills or ridges produced by hand labour, with long fallow periods between crops.

Farming in McLean County, Illinois, about a hundred miles south-west of Chicago in the heart of the US's Corn Belt, was changing from one tradition to another in the 1950s. It was a time when tractors were replacing horses, which meant that the oats and hay grown for the horses could be replaced by more corn and soya beans, and about half of the farms had mechanical corn pickers. But mechanical cultivators and hoes were still the only way of controlling weeds, and farms were family businesses. Several hundred miles to the west, however, traditional farming meant cattle ranching, with much bigger landholdings, whereas a few hundred miles south it meant small sharecroppers growing cotton.

In Europe, in the early 1950s, traditional farming meant mixed farming on family farms. On a typical farm in lowland England, for example, there might be a small dairy herd with Shorthorn, Ayrshire, or Jersey cows, with the calves that were not needed as herd replacements reared for beef, and a sheep flock. A few pigs would be found somewhere in the farm buildings, and chickens and geese would be wandering round the farmyard. Out in the fields there would be cereal crops, often a few acres of potatoes, and sometimes sugar beet. A good proportion of the land, perhaps a quarter to a half, would be in grass, producing summer grazing for the cattle and sheep and a crop of hay for winter fodder. There would be a few acres of root crops such as swedes, turnips, or mangel wurzels, also for winter fodder.

Mechanization would not have gone as far as it might have done in McLean County, but the cows would usually be machine milked into a bucket that was carried around the cowshed. Small tractors such as the little grey Ferguson were becoming more and more popular, and combine harvesters were beginning to replace the reaper-binder. But it is important to remember that this is the image of traditional farming in the mind of somebody who was young in the 1950s. Middle-aged farmers at the time would have thought of horses and hand milking as traditional, and tractors and milking machines as modernity. And if we go back a hundred years before they were born, to the early 19th century, we find John Clare, in his poem *The Mores*, bemoaning the process of enclosure that produced the patchwork fields of Midland England. He was used to open fields, worked in common, where

> Cows went and came, with evening morn and night,
> To the wild pasture as their common right.

And complained that

> Inclosure came and trampled on the grave
> Of labour's right and left the poor a slave.

There is, therefore, a frequent identification of farming with tradition, even if what is seen as traditional might change over time. Farming, in this view, is seen as a way of life, in which doing right by the land, producing healthy crops and livestock, employing local people, and having a thriving and well-maintained farm to hand on to the next generation are more important than expansion, profit maximization, and integration with the food chain. There is none of the division between work and home or labour and leisure that characterizes the alienated urban worker. The whole family is involved in the farm, and they work with both mind and body, maintaining a close relationship with place and the local community. There is independence and dignity in what they do. Farmers have a sense of identity, emerging from their shared concerns, their reading of the agricultural press and membership of farmers' organizations, their shared leisure pursuits, even their style of dress and selection of friends and marriage partners.

Clearly this is an idealized image, but it is a long-established and powerful one. It is a version of the pastoral, that long-established literary form, written from the perspective of the town or court, in which the country is something other, and the simple rural virtues of honesty, peace, and innocence are contrasted with the courtly virtues of wit and cunning.

Perhaps more accurately it is the georgic, in that it idealizes a country life as one of satisfying work rather than idleness. In the US it has a long history as 'Jeffersonian Agrarianism', based on the idea that farmers are the most valuable citizens, with ideal social values. It emerges from the way in which people remember the countryside of their youth, but it is reinforced by images in books, in films (think of the image of the farmhouse in the film *Babe*), in photographs, and in paintings (Constable's *The Hay Wain* is perhaps the classic example). It also ignores the negative aspects of the image, the suspicion of strangers and the high rates of suicide and industrial accidents. In contrast to big, modern, mechanized, globalized agribusinesses it is sustainable, produces wildlife habitats and beautiful landscapes, and cares for the welfare of animals. How accurate this contrast might be is something that we shall examine in the following parts of the chapter.

## Sustainability

The extent to which traditional farming in developed countries was sustainable is debatable. As we saw earlier, what counts as traditional depends not only on where you look, but when. The sort of farming that was carried out in North America and Europe in the middle of the 20th century, which present generations might now see as traditional, was already using fossil-fuel-powered tractors and fertilizers, although not to the extent to which they are used now. Tractors were beginning to make a significant impact from the 1930s, and from the middle of the 19th century some European farmers were using artificial fertilizers, especially phosphates.

The expansion of European animal production at the end of the 19th century was heavily reliant on cheap feedgrains imported from the US and Canada (which also supplied bread grains), so if we wish to go back to the point when Europe was feeding itself on entirely renewable resources we have to return to the middle of the 19th century, if not earlier. The population of Europe then was about 260 million, whereas today it is nearly three times as many.

In 1987, the World Commission on Environment and Development produced a report for the United Nations General Assembly, often called the Brundtland Report, from the Commission's chair, Gro Harlem Brundtland. It defined sustainability as meeting the needs of the present without compromising the ability of future generations to meet their own needs. Thus if the activities of our generation leave insufficient reserves of energy, water, or fertilizers for future generations, create global warming, pollute the environment, or destroy wildlife habitats and landscapes, they would be identified as unsustainable.

Clearly this has implications for agriculture. Different crops and animal enterprises have differing pollution effects, and, as we saw in the discussion of carbon footprints in Chapter 4, different crops and animal enterprises produce widely varying greenhouse gas emissions, implying that, from a climate change perspective, some are more sustainable than others.

But there is a further sustainability perspective too: will there be sufficient reserves of the finite resources that farming uses to cater for the needs of subsequent generations of farmers? This question applies to all kinds of inputs, from the soil and water that farmers have always used to the tractor fuel, electricity, fertilizers, and pesticides that are an integral part of modern farming in developed countries, and increasingly in developing countries, and it is controversial.

A critique of modern farming claims that a tonne of maize produced in the US for animal feed requires a barrel (42 US gallons or 159 litres) of oil, with two barrels of oil being needed to make the fertilizer and pesticides used on an average hectare of crops, so that agriculture is responsible for 7 per cent of the US's energy use. From a different perspective, another writer points out that agriculture accounts for only 3 per cent of global energy consumption, and argues that in times of scarcity oil tends to be reserved for vital uses, one of which would be food production.

Another important and finite resource is phosphorus. As we saw in Chapter 1, it is one of the three major constituents of the fertilizers that farmers use. Much of the total of 145 million tonnes that are mined each year goes into fertilizer, and at that rate the reserves that are worth exploiting at present prices and with present technology would, according to an estimate from the UN's Food and Agriculture Organization, be worked out in about 82 years. However, it is important to remember that the rate of use can vary considerably and that mining technology can change, so there are considerable disagreements about when the world will reach 'peak phosphorus'. Similarly with another principal fertilizer ingredient, potassium, for which the US Geological Survey calculates that the available reserves of 9.5 billion tonnes would suffice for 287 years at the present rate of consumption.

Nitrogen, the third major inorganic fertilizer constituent, is different. It is relatively easy to liquefy air and then distil it to separate out the nitrogen, but to be made into useful fertilizers such as ammonium nitrate or urea the nitrogen first needs to be combined with hydrogen to make ammonia. The usual feedstock for this process is natural gas, and making nitrogen fertilizers uses about 5 per cent of the annual global gas consumption.

Pesticides, being chemically much more complex and used in smaller total quantities, do not impose anything like the resource needs of fertilizers, but again, like fertilizers, have pollution implications that control the extent to which they can be used. One estimate of the cost of the damage caused by agriculture in the US to human health, water, soil, air quality, and wildlife suggested a range of figures between $US5.7 billion and $US16.9 billion in 2002, with a further $US3.7 billion in government expenditure on regulation and damage mitigation.

To be sustainable, therefore, either on a global or local scale, modern agriculture needs sufficient land of reasonable agricultural quality, with neither too much nor too little water on it, and a reliable supply of energy, fertilizers, and pesticides that can be used without producing major environmental problems. It is often argued that traditional farming methods are sustainable because they produce their own energy and fertilizers. The energy comes from the muscles of people or draught animals, both of which are fed by the food produced on the farm. The fertilizing nutrients are recycled in the farmyard manure produced by the animals (and often by the people too). Pests and diseases are controlled by selecting resistant varieties, mechanical weeding, fallowing, and crop rotation, or accepted as problems. Therefore, it is argued, this is the kind of farming that we must maintain where it still exists, or to which in the long run we must return. The contrary argument is that this kind of farming would produce too little food for the present world population. Already, according to one estimate, two-fifths of the world population are fed on food produced with the aid of nitrogen fertilizers.

Modern farming can, however, change to become more sustainable. One way of doing it is simply to adopt organic farming methods, in which pesticides and inorganic fertilizers are not used. Yields are lower, sometimes only half as much as conventional farming in the case of cereals, but very little lower in the case of dairy cows. Most organic farmers in developed countries still use tractors, so there is no saving in fossil fuel use. This form of farming usually accounts for only a small proportion of the land: in the United Kingdom, in 2004, for example, it was about 4 per cent of agricultural land.

But there are other approaches. One of the most dramatic responses to changing input availability has occurred in Cuba since 1990. In the 1980s, Cuban agriculture was dominated by large-scale sugar plantations, the produce of which was sold to the Soviet Union and other eastern bloc countries. It was heavily reliant on imported fertilizers, pesticides, oil, and cereals. With the break-up of the Soviet regime between 1989 and 1992, and a simultaneous trade embargo applied by the US, its main market disappeared and it could no longer afford its principal imports. Cuba's response was extraordinarily rapid. Within a few years the state farm sector was turned over to worker-run co-operatives operating on a much smaller scale, food crops and farm animals were increasingly found in the cities, farmers abandoned tractors for which there was little fuel and few spare parts and returned to animal traction, and synthetic fertilizers and pesticides were replaced by the use of compost and animal manures, resistant varieties, rotations, and intercropping, with predatory insects for pest control.

In a sense all this was simply a response to market forces: input prices rose and so did agricultural prices, so farmers had a greater incentive to produce food, but by using a different set of inputs. The government also played a part by changing the pattern of landholding, using Cuba's well-established network of agricultural scientists, and subsidizing food prices to consumers.

When insecticides first became available there was clearly a tendency to overuse them. In Sabah (Malaysia), in the 1960s, for example, the use of organochlorine insecticides on newly established cocoa crops affected both pest species and their natural predators, with the result that it was the pests that increased and destroyed the crop. After spraying was stopped, some of the pests were controlled using parasitic wasps and others by removing the trees that were their natural hosts.

There are numerous similar stories, and they have led farmers and agricultural scientists to realize that pests and pathogens can rapidly evolve resistance, that detailed study of their life histories can often suggest control methods, and that it is often better to live with low levels of pest attacks kept within bounds by a variety of control approaches than to attempt to wipe out the pest altogether.

In developed countries, too, increased fertilizer and pesticide prices, and changing agricultural policies, have encouraged farmers to find more sustainable farming methods. Instead of broadcasting granular fertilizer, some of which inevitably ends up in hedges and ditches, they spray liquid fertilizer, which can be placed more accurately. The imposition of Nitrate Vulnerable Zones in the UK requires farmers to spread fertilizer only at times of the year when crops are growing and so reduces losses to the water table. Those with smart phones can now download the Farm Crap App, which makes a visual assessment of manure and slurry application rates and calculates the nutrients provided, so producing potential savings in purchased artificial fertilizers.

Over 90 per cent of the UK cereal area is now sprayed with various herbicides, fungicides, and insecticides, and over half the area gets more than four spray treatments per year, but the total of active ingredients applied decreased from 15.5 million kg in 1990 to 9.4 million kg in 2010. These changes reflect the way in which society values what some might see as by-products of agriculture, in the form of pleasant landscapes and wildlife habitat, and the following section examines their relationship with farming in greater detail.

## Wildlife and landscape

The American biologist Rachel Carson was one of the first writers to draw attention to the potential impact of modern agriculture on wildlife. She envisaged a time when pesticides would have wiped out the insects and weed species on which birds live, so that the birds themselves would no longer be there to sing. Hence the title of her book, *Silent Spring*, which was published in 1962. Since then, numerous other writers have drawn attention to the impact of modern farming methods on the environment.

In tropical countries the main concern has been over the loss of rainforest, with its impact on global warming. According to a report by the UN Food and Agriculture Organization in 2011, 7.1 million hectares (out of a total of 1.3 billion) were lost each year from the forests of the Amazon, the Congo Basin, and South East Asia in the 1990s (Figure 7). The annual figure decreased to 5.4 million hectares in the 2000s, more than a million of which were for the expansion of soya cultivation. Much of the rest, in South America, was the result of the expansion of cattle pasture.

7. Crop land displacing the Amazon rain forest.

In temperate countries too, agriculture affects wildlife and its habitats, and also the landscape. Clearly these two are linked, since the agricultural landscape is a wildlife habitat. But it is quite possible for farmers to create pleasant landscapes, with a patchwork of grass fields and waving corn, which offer little food or shelter to wildlife. Equally, an area of impenetrable overgrown scrub, which most people would consider to have little landscape value, might offer both food and habitat to a variety of mammals, birds, and insects. The following discussion will therefore make what might appear to be a rather artificial distinction between temperate agriculture's impact on wildlife and landscape.

The idea of wildlife and wilderness is, as Roderick Nash pointed out in his classic book *Wilderness and the American Mind*, dependent upon the existence of agriculture. Without fields in which plants grew because they had been sown there by humans, the term 'wilderness' was meaningless; and animals were just animals until the domesticated ones could be distinguished from the wild. Thus agriculture affected nature from its very beginning, and although modern agriculture has clearly had a major impact on wildlife, it is worth remembering that there is nothing new in this.

In 1566, for example, the English Queen Elizabeth I's government passed 'An Acte for the Preservation of Grayne', which remained in force until 1863 and provided for bounties to be paid for the destruction of what were deemed to be pest species of birds and mammals: a penny for the head of a bullfinch or the heads of twelve starlings or three rats, a halfpenny for the head of a mole, and so on. On the other hand, the activities of farmers also provided wildlife habitats, from the artificial woodland edges provided by hedges to the roosts for owls and bats in old barns.

In the last fifty years or so, however, the impact of modern agricultural techniques in temperate countries has reduced the extent and variety of wildlife habitat, the variety of wild plants, and the numbers of animals. Whereas the grass grew long in traditional late-cut hay meadows and sheltered partridges, corn buntings, and corncrakes, modern silage swards are cut shorter and earlier and consist of grass and few of the flowers that flourished in traditional meadows. Modern machinery works best in big fields, so many miles of hedge have been grubbed out, with a consequent loss of nesting sites and flowers. The impact of organochlorine pesticides on the breeding success of birds of prey in the 1950s and 1960s is well known. Their numbers consequently fell significantly, both in Western Europe and North America, where the peregrine became extinct east of the Rocky Mountains. More recently, the neonicotinoid insecticides, which only became widely used in the 1990s, have been identified as the cause of decreasing numbers of honeybees and also of insect-eating birds such as the skylark. These are just some of the better-known examples.

Farming has always affected the appearance of the landscape, from the time that the first farmers cleared land for their crops.

The chequerboard patterns of farms and square fields that the airline passenger sees while flying over the Midwestern states of the US were the result of government decisions made in the 19th century. The smaller scale farming landscapes of Western Europe evolved more gradually, with medieval open fields being converted at different times to smaller enclosed fields, which in recent years have again grown in size, while often new factory-like buildings have been erected to accompany the traditionally styled farmhouses and barns.

In addition, it is important to recognize that farming activities also change the landscape in the short term. Ploughed arable fields turn green as cereal crops germinate and then change to waving gold as harvest approaches. Up to the middle of the 20th century, cereal fields could also be red as a result of poppies growing among the corn, but the increased use of selective herbicides has now virtually eliminated that colour from the landscape. Great expanses of green oilseed rape (canola) become a vivid yellow when the crop flowers in the spring or early summer and then subside to a dull brown as the plants set seed. Even the colour of the cattle in the fields can change as a result of a farmer's decision: the brown, or brown and white, Shorthorns and Ayrshires that were found in dairying districts of Britain sixty years ago have now been almost entirely replaced by black and white Friesians and Holsteins.

## Animal welfare

Animal welfare issues, like the questions of environmental sustainability discussed in the previous section, are easier to raise in developed economies where food is plentiful than in countries where food is scarce.

Keeping animals on farms implies that we accept the philosophical position that it is justifiable to keep other animals for the benefit of humans, and it is important to recognize that some people take the position that killing animals for human food is wrong, while others go further and argue that any form of animal exploitation, such as keeping them for their milk or wool, is unjustifiable. Clearly farmers who keep animals (and consumers who eat them and their products) do not take this position, but they would presumably all agree that abusing animals is wrong.

The more difficult question is to decide upon what constitutes abuse. Traditional small-scale farms, on which the farm staff are familiar with individual animals, should ensure the welfare of farm livestock; keeping large numbers of animals in close confinement is more likely to create physical and psychological deprivation and disease problems. This, at its simplest, is the way in which the potential animal welfare problem of modern farming is perceived, and stating it in this way suggests that there are different aspects of the problem: the extent to which the animal is prevented from exhibiting its normal behaviour; the effects of scale; and the question of disease.

Anyone who has seen animals who have been kept confined racing round the field into which they have been released before settling down to graze will recognize that they are more than biological machines. Seeing the apparent pleasure that a hen gets from a dust bath suggests that there might be more to her welfare than simply having enough to eat and drink. Conversely, animals kept on comfortable bedding and sheltered from cold, wind, and rain, or the blazing noonday sun in hotter climates, are presumably better off than those without such protection.

Traditional farming methods may still involve practices such as dehorning, castration, and branding which clearly interfere with the welfare of the animal, if only briefly, although practices such as debeaking chickens (i.e. cutting off the tips of their beaks so that they are less likely to peck each other) and docking the tails of pigs, both of which were designed to overcome intensive housing problems, are now much rarer. There are also degrees of confinement. Animals kept in groups in large pens are in a different position from a sow in a farrowing crate or hens in a battery cage, and advocates of housing point out that farrowing crates prevent sows from lying on their offspring, and that a stressed animal will be unproductive and prone to disease, so that no farmer will do anything deliberately to produce such stress.

This is where the second factor—the scale of operation—becomes significant. The level of expertise required to manage a large unit with hundreds or thousands of animals is of a different order of magnitude from that needed with traditional small flocks and herds, and it is much more difficult to give individual attention to each animal. As farmers have known for a long time, disease transmission is also easier when large numbers of animals are kept together. Hence old expressions of farming wisdom such as 'a sheep's worst enemy is another sheep'.

High productivity can also result in extra stress on an animal. A cow that is producing a lot of milk needs a big udder, which imposes extra strain on the attachment ligaments, and she will also be more prone to mastitis (an udder infection) and foot problems. Liver abscesses have been found in feedlot cattle fed on highly concentrated, low roughage diets. All these are problems for farmers, but they are also questions that need to be confronted by society in general when reflecting upon the kind of farming that we want to produce our food.

# Chapter 6
# Farming futures

Little so far has appeared in this book about government involvement in agriculture, but national governments as well as international organizations—such as the Food and Agriculture Organization of the United Nations and the World Trade Organization—all have an interest in agriculture. So too do numerous non-governmental organizations, pressure groups, and charities.

In this chapter we shall look at some of the most important questions that agricultural policymakers currently have to think about: what will be the impact of climate change; what is the impact of genetically modified organisms (GMOs) and are they safe to use; and will we be able to feed the world in the future? But first we need to look at some of the arguments around agricultural policy.

## Governments and farming

Most governments in industrialized countries support their agricultural industries in one way or another, but the extent to which they do so varies considerably. Whereas farmers in Australia and New Zealand get hardly anything, those in Norway, Iceland, Switzerland, Japan, and South Korea received between one-half and two-thirds of their receipts from government subsidies in 2011–13. Since in most cases agriculture forms only a small part of the economy of industrialized countries we might wonder why they support farmers, and why they often have government ministries with specific responsibility for agriculture, but not, for example, for retailing, which employs more people and contributes more to most developed economies. And since, in capitalist economies, markets are supposed to ensure the fulfilment of consumers' wants, why would governments in such societies want to intervene in agricultural markets at all?

Logically, the answer must be that agricultural markets do not produce what societies and governments want. They fail to do so in several different ways. First, they are not good at dealing with price fluctuations. The demand for agricultural products is fairly constant, because food consumers want to eat every day, and in roughly similar amounts, except perhaps at festivals such as Christmas. The supply of these products, however, can change considerably as a result of weather and disease changes. The result, as we saw in Chapter 3, is a fluctuation in price. High prices make life difficult for poorer consumers, who have to spend a high proportion of their incomes on food. Low prices may mean that marginal farms, which might survive in the long run at average price levels, are driven out of business by one or two low-income years. Consequently there has in the past been widespread agreement that dealing with such price fluctuations is worthwhile. The problem is that policies to control price fluctuations may become policies to raise farm prices and/or incomes, and these are more controversial.

This takes us on to other ways in which agricultural markets fail. One problem is that the firms to which farmers sell normally have much more market power—power to influence prices through their own decisions—than farmers, simply because they are bigger. Another is that the market does not normally pay for the ecosystem services, in the form of wildlife habitats and pleasant landscapes, that farmers deliver as a by-product of food production. Yet another is that the demand, in industrialized countries, for the basic food commodities that farmers produce does not increase very rapidly (see Chapter 3). Consequently, if farmers adopt new technologies and increase output, prices will tend to fall. In addition, they may face competition from farmers in other countries who can produce profitably at lower prices.

Farm lobby groups argue that governments in industrialized countries should tackle these issues. Their argument is that a country should be able to feed itself, or at least maintain a reasonable level of self-sufficiency, for several reasons.

One is national security. Governments have a responsibility to see that food is available in an emergency, although it might be argued that stable trading relationships are just as effective a form of food security, because food exporters are as concerned to ensure access to markets as importers are to secure supplies. Another argument concerns employment: a run-down agricultural sector means not only jobs lost in the countryside but also declining employment in the industries that supply farmers, such as fertilizers and farm machinery, and in the food industry that processes their products. There is also the question of paying for imported food. Governments have to decide whether their national resources are best employed in producing their own food or in producing something that can be sold in return for food.

Even if, on weighing up all these considerations, a government decides against intervening in agricultural markets, farmers might still argue that they are faced with declining and fluctuating incomes through no fault of their own. In the UK, the income per farmer in 2000 was only about a quarter of its 1976 level, although by 2010 it had recovered to just over half of that level. Since other groups in society (old-age pensioners and the unemployed, for example) receive state support in these circumstances, shouldn't farmers receive it too? On the other hand, there are different social groups, such as small shopkeepers and steelworkers, who work in traditional or declining industries and receive little or no state assistance, and who, unlike owner-occupying farmers, do not have a large capital asset such as a farm to fall back on.

These ideas should make it clear that questions of farm income-support involve value judgements which societies normally resolve through political mechanisms. Thus the function of political pressure groups, such as the National Farmers' Union (NFU) in the UK, or the American Farm Bureau Federation, the National Grange, and the National Farmers Union in the US, or the Comité des Organisations Professionnelles Agricoles (COPA) in the EU, is to put the farmers' case. Other groups, such as those representing the food industry, or environmentalists, may argue differently.

The history of agricultural policy in industrialized countries suggests that governments respond both to effective lobbying and to changing economic and political circumstances. In the US, intervention in agricultural markets can be traced back to the Agricultural Adjustment Acts of the 1930s, which not only allowed the federal authorities to buy grain from farmers, but also to pay them for not growing food on part of their land. This was the policy ridiculed by Joseph Heller in his novel *Catch-22*, in which Major Major's father was paid for not growing alfalfa, so he didn't grow more, and with the money the government paid him he bought more land upon which he didn't grow still more.

US agricultural policy is still concerned with evening out market fluctuations. Among the provisions of the Agricultural Act of 2014 farmers can be paid various sums from the public purse when market prices are low, or, in the case of dairy farmers, when the gap between milk prices and feed costs falls below a certain trigger level, at which point the Secretary of Agriculture can instigate purchases of dairy products to be used in domestic food programmes.

Successive UK governments in the late 19th and early 20th centuries relied on imports of agricultural products, especially from the Americas and Oceania, until the late 1930s. Then sea-borne trade, which at that time accounted for nearly two-thirds of UK food supplies, came under threat from surface warships and submarines. In response, the government initiated a series of price subsidies to promote increased domestic agricultural output. These were maintained after World War II because the UK economy was no longer in a position to pay for the previous level of imports, and were successful in raising the domestic self-sufficiency level to roughly two-thirds of domestic consumption by 1973.

The UK then joined what became the EU, which had a Common Agricultural Policy (CAP) operating on a different basis. It maintained domestic prices at higher than world levels using import tariffs, while surplus produce was initially bought into official stores and subsequently sold to the world market with the aid of export subsidies. This policy was very successful in doing what it was initially designed to do, which was to increase European agricultural output. But by the 1980s it was attracting considerable criticism as a result of its cost, its environmental impact, and its disruption of world agricultural markets. In consequence, it has undergone radical change.

Barriers to importing into the EU, which is the world's largest agricultural importer by a big margin, have been reduced, and farmers' incomes are now supported by direct payments. These do not encourage them to produce more than can be sold on domestic and export markets, but they are dependent upon the size of the farm business, so that big farmers receive more money than small farmers. This leads, in some cases, to very rich people receiving money from the public purse; whether this is justified by the argument that it represents a payment for the environmental services they provide is for the reader to decide.

In developing countries and emerging economies the variety of agricultural policies is as great, if not greater, than in industrialized countries, and has included measures to raise or stabilize prices, input subsidies, investment in education and technical training, research and development, advisory work, development of credit institutions and producer co-operatives, and land reform.

Part of the problem is that the opinions of experts and politicians on the most effective way of promoting development have changed over time. Fifty years or so ago it was thought that agricultural development was a prerequisite for development of the economy as a whole. Then it appeared that investment in education and health paid off more quickly, so in several countries farm prices were kept low and investment concentrated in urban areas. More recently opinion has swung back towards promotion of agriculture. At the same time, opinions on the best ways of doing so have changed. Whereas in the 1960s the emphasis was on research and development and subsidies for seeds and fertilizers, it later changed to attempts to identify the most appropriate price levels, but recently renewed attention has been given to R&D and the World Bank has advocated seed and fertilizer subsidies.

## Climate change

The reality of climate change is now widely accepted, although a survey of 1,276 farmers in Iowa in 2011 found that 32 per cent of them thought that climate change was not occurring, or that there was insufficient evidence of it. The International Panel on Climate Change (IPCC) has predicted a global average temperature rise of between 2°C and 4°C above that of the pre-industrial climate by 2050. To put it another way, global average temperatures rose by about 0.13°C per decade between 1950 and 2010 and are projected to rise by about 0.2°C per decade in the next three decades.

These increased temperatures are likely to increase the level of activity in the hydrological cycle, which in simple terms implies more rain. Where it will fall and when is more difficult to predict, especially in monsoon areas, although changes in the timing and pattern of seasonal rainfall seem likely. Higher temperatures also affect polar ice caps and mountain snow melt, leading to a predicted rise in sea level of two metres by 2100. It is also possible for climatic change in one region to have impacts elsewhere, the most obvious example being that less snow in the Himalayas implies more drought on the plains around the Indus and Ganges rivers. In addition to these gradual changes, there are also likely to be increases in climatic variability and extreme weather events, meaning longer and more frequent droughts, an increased risk of extreme wet weather leading to floods in the northern hemisphere, and fewer but longer and more intense tropical cyclones.

## Agriculture

Agriculture and climate change have a twofold relationship. Climate has an impact on agriculture, and agriculture has an impact on climate. Taking the latter first, we have already discussed (in Chapter 4) agriculture's carbon footprint, pointing out that agriculture makes a significant contribution to the world's greenhouse gas emissions, even if its precise extent is disputed. A high proportion—one estimate is 75 per cent—of these emissions originate in low- and middle-income countries, where rice production and biomass burning are concentrated. Consequently farmers are under pressure to change, both in response to policymakers and in their own self-interest. They may need to upgrade their drainage systems to cope with increased rainfall, or adapt their cropping to shorter growing seasons or more frequent droughts. In order to reduce greenhouse gas emissions they could increase soil carbon by changing the crops they grow or the cultivation methods they use. They could also process animal and crop wastes in anaerobic digesters to collect methane instead of allowing it to escape to the atmosphere, and increase the use of organic manures in order to reduce their fertilizer consumption.

As far as the impact on agriculture is concerned, perhaps the most dramatic of the expected effects of global warming, apart from catastrophic storms and floods, is the loss of agricultural land as a result of rising sea levels. In coastal zones, and especially in the river deltas of Bangladesh, Myanmar, India, Pakistan, and Egypt, one estimate suggests that some 650 million people could be affected. Conversely, reduced snowfall in the Himalayas is predicted to produce water shortages in the dry season many miles away in the basins of the Indus, Ganges, and Brahmaputra rivers, relied upon by 500 million people for domestic and agricultural purposes. In Africa, the amount of water in the Niger river basin decreased in the 20th century, and the drought in Russia in 2010 reduced wheat yields by 40 per cent.

Increased temperatures and changing rainfall patterns can restrict the length of the growing season. In northern Ghana, for example, farmers remember the rainy season being long enough to get two harvests, whereas now the second crop is at risk if the rains fail. Over the world as a whole, about 80 per cent of the agricultural area is rain-fed (the proportion is higher in Africa and Latin America), but the remaining 20 per cent of irrigated land produces over 40 per cent of the world's food, and is thought to be vulnerable to climate-change-induced water shortages.

Temperature fluctuations also affect crop yields, especially high temperatures (above 32°C) during the flowering stage of cereals. Above this temperature rice yields have been shown experimentally to be reduced by 90 per cent. Every degree-day above 30°C in maize (corn) crop trials reduced the final yield by 1 per cent under optimal rainfall conditions and 1.7 per cent in a drought.

However, it is important to remember that higher temperatures are likely to be beneficial in temperate regions, since the potential crop-producing area could increase, as will the length of the growing season, and possibly rainfall, while cold periods decrease.

Higher carbon dioxide levels in the atmosphere also have a fertilizing effect on crop growth (see Chapter 1). Consequently agriculture in North America, southern Latin America, and Europe is less likely to be adversely affected than sub-Saharan Africa and Asia, and could see yield increases in some crops.

Conversely, warmer temperatures also affect the ability of pests to survive and thrive. Aphids and weevil larvae have been shown to grow faster in higher carbon dioxide concentrations, and warmer winter temperatures would reduce seasonal mortality, allowing earlier and greater dispersion of pest species. Animal health is also likely to be affected. Perhaps the best-known recent example is in the spread of blue tongue disease in cattle and sheep to north and west Europe. The blue tongue virus is transmitted by the bite of midges of the *Culicoides* genus, and has been recognized for a long time around the Mediterranean, with *C.imicola* as the insect vector. It appears that the virus has recently been acquired by other midge species such as *C.obsoletus* and *C.pulicaris*, and their ability, and that of the virus, to survive has been increased by milder winters. Hence the spread of the disease, which reached south-east England by 2007.

Given the wide range of possible effects, it is not easy to predict the overall impact of climate change. A study of temperature and precipitation effects between 1980 and 2008 concluded that temperature had had a bigger impact than precipitation, and that maize production was 3.8 per cent lower, and wheat 5.5 per cent lower, than either would have been in the absence of the climate trends in that period. Gains and losses in rice and soya bean production roughly evened out, with rice output in temperate areas benefiting from higher temperatures. There could be big differences between countries: wheat yields in Russia were reduced by almost 15 per cent, whereas there were no effects in the US because there were no significant climate trends there. The authors of the study admitted that their model was likely to be pessimistic because it neglected the effects of carbon dioxide fertilization. Taking these into account, they estimated the total increase in average commodity prices resulting from climate change over these twenty-eight years at 6.4 per cent.

Using studies like this in attempting to predict the future is, however, fraught with difficulty. Not only do they neglect the possibility of farmers adapting to changing circumstances, they also neglect changes in the wider world within which farmers will be living and working. Hence the IPCC's idea of a Climate Vulnerability Index, which combines seventeen different variables, not only climatic but also demographic, socio-economic, and measures of governance. This clearly identifies much of Africa and Asia as most vulnerable, and North America and most of Europe as least vulnerable.

## GMOs

Traditional breeders worked by attempting to combine the desirable traits of different varieties or breeds of the same species. Early in the 20th century, for example, Professor Biffen at Cambridge crossed an established English wheat variety called Squareheads Master with a Russian variety, Ghirka, which was resistant to yellow rust, a fungal disease of the foliage of the wheat plant, to produce a new rust-resistant variety called Little Joss. Similarly, animal breeders have crossed a sheep breed with a high milk yield with another breed with a high lambing percentage in order to produce breeding ewes capable of producing plenty of milk for twin lambs. Hence the importance of rare breeds of livestock and exotic plant varieties. They represent a reservoir of genes with different properties that are available for future use in response to unpredictable changes in requirements.

The problem with these traditional approaches is that they are time consuming, so it is expensive to produce new breeds or varieties. Work on producing a potato variety resistant to attack by the fungus causing blight (see Chapter 1) began at the Scottish Plant Breeding Station (now part of the James Hutton Institute) in 1937. The variety Pentland Ace, which incorporated genes from a blight-resistant Mexican potato, was not released until 1951, and this is said to be an example of a relatively rapid breeding programme. Moreover, there are limits to these conventional methods: if a trait (e.g. blight resistance) is not present somewhere in the genes of a species it cannot be invented. From the 1920s crop breeders tried to increase genetic variation by artificially inducing mutations using, for example, X-rays or gamma rays. This is a random technique requiring very large numbers of plants to produce a few useful individuals, but it has been used to produce new varieties of oilseed rape (canola), rice, and barley.

Genetic modification offers, in theory at least, a more rapid and precise way of incorporating new traits into crops. Several stages are involved. First an organism must be found with the desired characteristics. The second stage is the isolation of the gene responsible for producing the desirable trait in the donor organism, which is done by using restriction enzymes to cut its DNA at the desired location. This gene is then cloned—copied many times—before being inserted into the target plant using a 'gene gun' or a bacterium. Using a gene gun means attaching the cloned genes to tiny metallic particles which are then inserted into the target plant cells under high pressure so that the donor DNA is integrated into the DNA of the target plant inside its cell nuclei. It sounds rather crude, but it has been used successfully, especially for monocotyledon crops such as wheat or maize.

For dicotyledonous plants such as potatoes, tomatoes, or carrots, the preferred technique is to use a bacterium, *Agrobacterium tumifaciens*, a natural plant parasite, to transport the donor DNA into the target plant. It's then a matter of testing the resulting plants to check that they have the desired characteristics, multiplying them, and putting them on the market.

One of the early, and still important, uses of the technique was to produce a soya bean variety that was resistant to glyphosate, a herbicide that kills nearly all green plants and is sold under the trade name RoundUp. Researchers working for Monsanto, the firm which produced glyphosate, found a bacterium that was resistant to it, and incorporated the gene responsible into a soya bean variety that they named RoundUp Ready. By the mid-1990s it was on the market, enabling farmers to control weeds by spraying with glyphosate without affecting the crop (and also increasing glyphosate sales).

Another well-known example was the ability of the soil bacterium *Bacillus thuringiensis* (Bt) to produce a protein toxic to insect larvae by affecting the cell membranes in their guts. Different strains of the bacterium produce varieties of the protein that affect different types of insects. Again the relevant genes could be identified. They were incorporated into several maize and cotton varieties by the mid-1990s and provided protection against the European corn borer, a major pest of maize, and the cotton bollworm and pink bollworm, major pests of cotton. Subsequent developments have resulted in the 'stacking' of traits, meaning that there are now maize varieties that have both Bt and glyphosate-resistant genes.

Genetic modification has also been applied, with varying degrees of success, to producing resistance to drought and to virus attacks. Rice varieties (so-called 'golden rice') have been engineered to have an increased vitamin A content and so help to protect against blindness. There are soya bean varieties with higher concentrations of omega-3 fatty acids, which are thought to help reduce cardiovascular disease. More experimentally, it has been used to produce transgenic farm animals. For example, genes from spinach and nematodes have been incorporated into pigs and dairy cattle to increase the level of omega-3 fatty acids in meat and milk, in order to reduce the risks of cardiovascular disease in people. So far, however, GM farm animals have remained within the laboratory.

In the twenty years or so since they were first introduced, the area sown to GM crops has increased enormously. A GM advocate claims that it has increased from 1.7 million hectares in 1996 to 170 million hectares across the world in 2012, evenly split between industrialized and developing countries. Much of this area is in the four crops that were involved from the beginning: soya beans, for which three-quarters of world production is now from GM varieties, cotton (nearly half GM), maize, and oilseed rape (in each case over a quarter GM). In some countries, such as the US, the proportion is higher, with most of the soya beans and rapeseed, and much of the cotton and maize, now coming from GM varieties.

There has also been an expansion in the range of crops involved, with GM versions of alfalfa, apples, courgettes, sugar beet, sugar cane, and wheat now available. Most of the Hawaiian papaya crop now comes from a variety that has had a gene added to produce resistance to the papaya ringspot virus.

Nevertheless, GM crops remain controversial, with most European countries banning, or carefully controlling, their use. Opponents of GM argue that combining modern scientific techniques such as genome sequencing and gene marker mapping, coupled with conventional breeding methods, are as useful as genetic engineering as far as increasing yields are concerned.

There is insufficient space in this book to go into all the arguments about GM in detail; listing them will have to do. There are fears that some people may develop allergies to GM varieties or find them toxic. Perhaps more significant are worries that modified genes might spread unintentionally, with unpredictable effects on non-target organisms. A study of GM maize being grown in Germany found that it was difficult to prevent wind-blown GM maize pollen from fertilizing non-GM maize plants. Similar studies in South Africa and Ghana led to concerns about the maintenance of traditional maize varieties.

It is also important to remember that plants and animals naturally vary in their susceptibility to toxicity, so that killing the most susceptible advantages those least so, and thus resistance will tend to build up in plant and animal species. Reuters reported in September 2011 that twenty-one glyphosate-resistant weed species, including marestail in Ohio and waterhemp in north-east Kansas, affected 11 million acres in the US. In order to control the development of resistance among pest species, farmers in the US are required to plant part of their crop areas with non-Bt varieties to act as refuges for non-resistant insects, but policing this policy would probably be difficult in other parts of the world. The Monsanto website blamed the breakdown of pink bollworm resistance in one of its Bt cotton varieties in India on the failure of farmers there to plant non-Bt refuge crops.

There are also social and ethical concerns. Since GM varieties normally have to be purchased afresh from the seed companies each year, the traditional practice of saving seed is no longer available, which can affect poor farmers in particular while benefiting the multinational companies that produce them. The effect that these actual or potential problems will have on the current expansion of GM crops remains to be seen.

## Feeding nine billion people

In 2010 the UN Population Division's medium estimate predicted that there would be 9.3 billion people in the world in 2050. In 2013 a *New York Times* article calculated that agriculture currently produces 2,700 calories for each of the 7 billion inhabitants of the world. If there were to be no increase in food output, current production would still produce just over 2,000 calories per person for the 2050 population, which is perfectly adequate for survival. So is there anything to worry about?

If you've read this far you will have realized that it's not quite as simple as that. About a third of those calories currently go to feed animals, some will be wasted as they go through the food chain, and at present about 5 per cent will be converted into biofuels. The remaining supplies are not evenly distributed, because the industrialized countries get more per person than developing countries, and richer people everywhere can spend more on food than poorer people. Not only that, but people in the future may have different food demands, and the resources and technologies available to produce food will be different. This book, let alone the remains of this chapter, is far too short to produce a considered estimate of the likely world food balance in 2050, but we can at least look at some of the things to think about in doing so.

As we saw in Chapter 3, the two main factors affecting future food demand are the number of people in the world and the amount of money they have to spend. That medium population estimate of 9.3 billion assumes that the average woman will have 2.17 children in 2045–50. In 2005–10 she had 2.52 children. But if she has 2.64 children, the high estimate, there could be 10.6 billion mouths to feed. Factors affecting female fertility, such as education, income, access to birth control, religion, and culture are therefore crucial as far as future food demand is concerned. On the whole, as societies become richer, people tend to have fewer children. But they also eat more, and have a more varied diet, as we also saw in Chapter 3. We have already seen an increase in demand for pork, and consequently in demand for feed grains, as the Chinese economy has grown, and we would expect other emerging economies to increase their demand for animal protein and fats as their consumers spend their increased incomes. In short, we should expect the demand for food to grow by more than the growth in population, but by how much more is considerably more difficult to predict.

What we do know is that in the past increasing demand, and especially increasing population, has been more than met by increases in supply. The historian Giovanni Federico calculated that while the world population rose six or sevenfold between

1800 and 2000, world agricultural production rose by at least ten times in the same period. In the 19th century most of the increase came from land in North and South America, and Australia and New Zealand, that had not previously produced food and fibre for the world market. In the 20th century the increase was mainly the result of increases in output from each hectare of land and/or person working on it. Can either of these solutions be applied again in the 21st century?

We are still cutting down trees in countries such as Argentina and Brazil in order to grow maize and soya beans, but this is increasingly coming under political pressure, and, as we saw earlier, we are likely to lose land as a result of sea-level rises, droughts, and desertification as global warming continues. It is not easy to predict how these trends will balance out in detail, but it does seem clear that we cannot expect anything like the sort of extra-land-based production increase that we saw in the 19th century.

That leaves us with producing more per hectare of agricultural land. In the past, faced with an increase of demand relative to supply, and consequently increased prices, farmers have been good at producing more by using extra inputs—from more hand labour in weeding to extra fertilizers and pesticides. Scientists and technologists have also been good at producing better plant varieties, livestock breeds, machinery, and so on. While world population increased by just over 80 per cent between 1970 and 2009, world maize production increased three times and wheat and rice production more than doubled. Can this continue into the future?

In industrialized countries, the scope for easy ways to increase output may be limited, but there already exists a highly trained and sophisticated agricultural labour force that is used to adopting new technologies such as integrated pest management in their search for the desirable realization of sustainable intensification.

The greatest opportunities for increasing output per hectare are also where the increase in demand is likely to be greatest: in the emerging and developing economies. They are the ones in which a little extra fertilizer, or slightly better seeds, or a little extra draught power from cattle or donkeys, can make the greatest difference. Farmers can use mobile phones both to receive technical advice and also to find out which local markets have the best prices. Even simple things such as a nearby piped water supply, or a way of cooking food that does not require a woman to spend two or three hours a day searching for firewood, could mean that more time is available for weeding a crop or growing a second crop or looking after animals. Bringing the world average yields of crops and livestock up to the levels already being achieved on the more productive farms is therefore the challenge for the 21st century.

# Further reading

## General

N. De Pulford and J. Hitchens, *Behind the Hedge: Where Our Food Comes From*, Ammanford: Sigma Press, 2012.

C. Martiin, *The World of Agricultural Economics: An Introduction*, London: Routledge, 2013.

E. Millstone and T. Lang, *The Atlas of Food: Who Eats What, Where, and Why*, Berkeley: University of California Press, revised edn., 2008.

R. J. Soffe (ed.), *The Agricultural Notebook*, 20th edn, Oxford: Blackwell, 2003.

C. C. Webster and P. N. Wilson, *Agriculture in the Tropics*, 2nd edn, London: Longman, 1980.

## On the history of agriculture

D. B. Danbom, *Born in the Country: A History of Rural America*, 2nd edn, Baltimore, MD: Johns Hopkins University Press, 2006.

M. B. Tauger, *Agriculture in World History*, London: Routledge, 2011.

J. Thirsk, *Alternative Agriculture: A History, from the Black Death to the Present Day*, Oxford: Oxford University Press, 1997.

## Chapter 1: Soils and crops

### On soils

R. J. Buresh, P. A. Sanchez, and F. Calhoun (eds), *Replenishing Soil Fertility in Africa*, Madison, WI: Soil Science Society of America, 1997.

E. A. FitzPatrick, *An Introduction to Soil Science*, 2nd edn, Harlow: Longman, 1986.

T. R. Paton, G. S. Humphreys, and P. B. Mitchell, *Soils: A New Global View*, London: UCL Press, 1995.

C. Reij, I. Scoones, and C. Toulmin (eds), *Sustaining the Soil: Indigenous Soil and Water Conservation in Africa*, London: Earthscan, 1996.

E. J. Russell, *Russell's Soil Conditions and Plant Growth* (ed. A. Wild), 11th edn, Harlow: Longman, 1988.

### On crops

J. Christopher, *The Death of Grass*, London: Michael Joseph, 1956.

H. J. S. Finch, A. M. Samuel, and G. P. F. Lane, *Lockhart and Wiseman's Crop Husbandry*, 8th edn, Cambridge: Woodhead Publishing, 2002.

S. Meakin, *Crops for Industry: A Practical Guide to Non-Food and Oilseed Agriculture*, Marlborough: Crowood Press, 2007.

J. W. Purseglove, *Tropical Crops: Dicotyledons*, vols 1 and 2, London: Longmans, 1968.

J. W. Purseglove, *Tropical Crops: Monocotyledons*, London: Longmans, 1972.

N. W. Simmonds and J. Smartt, *Principles of Crop Improvement*, 2nd edn, Oxford: Blackwell, 1999.

W. D. Walters, *The Heart of the Cornbelt: An Illustrated History of Corn Farming in McLean County*, Bloomington, IL: McLean County Historical Society, 1997.

## Chapter 2: Farm animals

T. G. Field and R. E. Taylor, *Scientific Farm Animal Production: An Introduction to Animal Science*, 9th edn, Upper Saddle River, NJ: Prentice Hall, 2008.

R. D. Frandson and T. L. Spurgeon, *Anatomy and Physiology of Farm Animals*, Philadelphia, PA: Lea and Febiger, 1992.

P. McDonald, R. A. Edwards, J. F. D. Greenhalgh, C. A. Morgan, L. A. Sinclair, and R. G. Wilkinson, *Animal Nutrition*, 7th edn, Harlow: Pearson Education, 2011.

M. L. Ryder, *Sheep and Man*, London: Duckworth, 1983.

R. T. Wilson, *The Camel*, Longman, 1984.

# Chapter 3: Agricultural products and trade

G. L. Cramer, C. W. Jensen, and D. D. Southgate, *Agricultural Economics and Agribusiness*, 8th edn, New York: Wiley, 2001.

O. Ecker and M. Qaim, 'Income and Price Elasticities of Food Demand and Nutrient Consumption in Malawi'<http://ageconsearch.umn.edu/bitstream/6349/2/451037.pdf>.

A. Regmi, M. S. Deepak, J. L. Seale Jr, and Jason Bernstein, 'Cross-Country Analysis of Food Consumption Patterns' <http://www.ers.usda.gov/media/293593/wrs011d_1_.pdf> (USDA elasticities data).

R. Tiffin, K. Balcombe, M. Salois, and A. Kehlbacher, 'Estimating Food and Drink Elasticities' <https://www.gov.uk/government/uploads/system/uploads/attachment_data/file/137726/defra-stats-food-farm-food-price-elasticities-120208.pdf>.

WTO, 'Merchandise Trade by Product' <http://www.wto.org/english/res_e/statis_e/its2013_e/its13_merch_trade_product_e.pdf>.

The website of the UN Food and Agriculture Organization, <http://faostat.fao.org/>, is an excellent source of global agricultural statistics.

# Chapter 4: Inputs into agriculture

D. Blandford, I. Gaasland, and E. Vårdal, 'Extensification versus Intensification in Reducing Greenhouse Gas Emissions in Agriculture: Insights from Norway', *Eurochoices* 12(3), 2013, pp. 4–8.

FAO, *FAO Statistical Yearbook 2013: World Food and Agriculture* <http://www.fao.org/docrep/018/i3107e/i3107e00.htm.

S. K. Lowder, J. Skoet, and S. Singh, 'What Do We Really Know about the Number and Distribution of Farms and Family Farms Worldwide?', Background paper for *The State of Food and Agriculture 2014*, ESA Working Paper No. 14–02, Rome: FAO, 2014.

M. M. Mekonnen and A. Y. Hoekstra, 'The Green, Blue and Grey Water Footprint of Crops and Derived Crop Products', *Hydrology and Earth System Sciences* 15, 2011, pp. 1577–600.

M. M. Mekonnen and A. Y. Hoekstra, 'A Global Assessment of the Water Footprint of Farm Animal Products', *Ecosystems* 15, 2012, pp. 401–15.

P. Scarborough, P. N. Appleby, A. Mizdrak, A. D. M. Briggs, R. C. Travis, K. E. Bradbury, and T. J. Key, 'Dietary Greenhouse Gas Emissions of Meat-Eaters, Fish-Eaters, Vegetarians and Vegans in the UK', *Climate Change* 125, 2014, pp. 179–92.

## Chapter 5: Modern and traditional farming

P. Brassley, 'Agricultural Technology and the Ephemeral Landscape', in D. E. Nye (ed.), *Technologies of Landscape: From Reaping to Recycling*, Amherst: University of Massachusetts Press, 1999, pp. 21–39.

D. Briggs and F. Courtney, *Agriculture and Environment: The Physical Geography of Temperate Agricultural Systems*, Harlow: Longman, 1989.

J. Burchardt and P. Conford, *The Contested Countryside: Rural Politics and Land Controversy in Modern Britain*, London: I.B.Tauris, 2008.

G. Conway, 'The Doubly Green Revolution', in Pretty (2005) pp. 115–27.

FAO and International Tropical Timber Organisation, *The State of Forests in the Amazon Basin, Congo Basin, and South-East Asia*, Rome: FAO, 2011.

R. Gasson and A. Errington, *The Farm Family Business*, Wallingford: CAB International, 1993.

C. A. Hallmann et al., 'Declines in Insectivorous Birds are Associated with High Neonicotinoid Concentrations', *Nature* 511, 17 July 2014, pp. 341–3.

R. Lovegrove, *Silent Fields: The Long Decline of a Nation's Wildlife*, Oxford: Oxford University Press, 2007.

P. Lymbery and I. Oakeshott, *Farmageddon: The True Cost of Cheap Meat*, London: Bloomsbury, 2014.

P. McMahon, *Feeding Frenzy: The New Politics of Food*, London: Profile Books, 2013.

R. Nash, *Wilderness and the American Mind*, 3rd edn, New Haven, CT: Yale University Press, 1982.

J. Pretty (ed.), *The Earthscan Reader in Sustainable Agriculture*, London: Earthscan, 2005.

P. Rosset and M. Bourque, 'Lessons of Cuban Resistance', in Pretty (2005) pp. 362–8.

V. Smil, *Enriching the Earth: Fritz Haber, Carl Bosch, and the Transformation of World Food Production*, Cambridge, MA: MIT Press, 2001.

E. M. Tegtmeier and M. D. Duffy, 'External Costs of Agricultural Production in the United States', in Pretty (2005) pp. 64–89.

W. D. Walters, *The Heart of the Cornbelt: An Illustrated History of Corn Farming in McLean County*, Bloomington, IL: McLean County Historical Society, 1997.

R. Williams, *The Country and the City*, London: Chatto and Windus, 1973.

UK Pesticide use statistics at <http://pusstats.fera.defra.gov.uk>.

For phosphate reserves: <http://www.fao.org/docrep/007/y5053e/y5053e07.htm> (consulted 19.1.2015).

For potassium reserves: <http://minerals.usgs.gov/minerals/pubs/commodity/potash/mcs-2011-potas.pdf> (consulted 19.1.2015).

## Chapter 6: Farming futures

### On agricultural policy

J. Brooks, 'Agricultural Policy Choices in Developing Countries' <http://economics.ucr.edu/seminars_colloquia/2010/applied_economics/Brooks%20paper%20for%201%204%2010.pdf>.

P. de Castro, 'A Comparative Approach to European and American Agricultural Policies', *momagri.org* <http://www.momagri.org/UK/points-of-view/A-comparative-approach-to-European-and-American-agricultural-policies_798.html>.

European Commission, 'The Common Agricultural Policy after 2013' <http://ec.europa.eu/agriculture/cap-post-2013/index_en.htm>.

European Commission CAP overview <http://ec.europa.eu/agriculture/cap-overview/2014_en.pdf>.

European Commission, *Agriculture in the European Union Statistical and Economic Information: Report 2013* <http://ec.europa.eu/agriculture/statistics/agricultural/2013/pdf/overview_en.pdf>.

P. Shelton, 'Can the US Farm Bill and EU Common Agricultural Policy Address 21st Century Global Food Security?', *IFPRI Blog*, 23 July 2014<http://www.ifpri.org/blog/can-us-farm-bill-and-eu-common-agricultural-policy-address-21st-century-global-food-security>.

## On climate change

Anon., *Climate Impacts on Food Security and Nutrition: A Review of Existing Knowledge*, Exeter: Met Office and UN World Food Programme, 2012.

D. B. Lobell, W. Schlenker, and J. Costa-Roberts, 'Climate Trends and Global Crop Production Since 1980', *Science* 333, 29 July 2011, pp. 616–20.

L. Nijs, *The Handbook of Global Agricultural Markets: The Business and Finance of Land, Water and Soft Commodities*, Basingstoke: Palgrave Macmillan, 2014, chapter 3.

C. L. Walthall et al., *Climate Change and Agriculture in the United States: Effects and Adaptation*, Washington, DC: USDA Technical Bulletin 1935, 2013, available at <http://www.usda.gov/oce/climate_change/effects.htm>.

## On GM crops

R. Boyle, 'How to Genetically Modify a Seed, Step By Step', *Popular Science* <http://www.popsci.com/science/article/2011-01/life-cycle-genetically-modified-seed?single-page-view=true>.

F. Forabosco et al., 'Genetically Modified Farm Animals and Fish in Agriculture: A Review', *Livestock Science* 153 (May 2013) pp. 1–9.

N. Halford, *Plant Biotechnology*, London: Wiley, 2006.

C. James, 'Global Status of Commercialized Biotech/GM Crops: 2012', ISAAA website <http://www.isaaa.org/resources/publications/briefs/44/highlights/default.asp>.

## On feeding nine billion

G. Conway, *One Billion Hungry: Can We Feed the World?* Ithaca, NY: Cornell University Press, 2012.

G. Federico, *Feeding the World: An Economic History of Agriculture, 1800–2000*, Princeton, NJ: Princeton University Press, 2005.

C. Juma, *The New Harvest: Agricultural Innovation in Africa*, Oxford: Oxford University Press, 2011.

P. McMahon, *Feeding Frenzy: The New Politics of Food*, London: Profile Books, 2014.

# Index

Index

Agriculture

# Expand your collection of
# VERY SHORT INTRODUCTIONS

# SOCIAL MEDIA
# Very Short Introduction

# Join our community

www.oup.com/vsi

- Join us online at the official Very Short Introductions
  **Facebook** page.
- Access the thoughts and musings of our authors with our
  online **blog**.
- Sign up for our monthly **e-newsletter** to receive information
  on all new titles publishing that month.
- Browse the full range of Very Short Introductions online.
- Read **extracts** from the Introductions for free.
- Visit our library of **Reading Guides**. These guides, written by our
  expert authors will help you to question again, why you think
  what you think.
- If you are a teacher or lecturer you can order inspection
  copies quickly and simply via our website.

# ONLINE CATALOGUE
## A Very Short Introduction

Our online catalogue is designed to make it easy to find your ideal Very Short Introduction. View the entire collection by subject area, watch author videos, read sample chapters, and download reading guides.